1994

W9-ACF-970

TECHNOLOGY ASSESSMENT IN SOFTWARE APPLICATIONS

TECHNOLOGY ASSESSMENT IN SOFTWARE APPLICATIONS

Edited by

HAROLD F. O'NEIL, JR.
University of Southern California

EVA L. BAKER
University of California, Los Angeles

 LAWRENCE ERLBAUM ASSOCIATES, PUBLISHERS
1994   Hillsdale, New Jersey   Hove, UK

Lawrence Erlbaum Associates, Inc., Publishers
365 Broadway
Hillsdale, New Jersey 07642

Library of Congress Cataloging-in-Publication Data

Technology assessment in software applications / edited by Harold F.
O'Neil, Jr., Eva L. Baker.
    p.   cm.
    Presented at the conference held at the University of California,
Los Angeles.
    Includes bibliographical references and index.
    ISBN 0-8058-1248-2 (cloth). — ISBN 0-8058-1249-0 (paper)
    1. Application software—Congresses. 2. Technology assessment—
Congresses. I. O'Neil, Harold F., 1943–   . II. Baker, Eva L.
QA76.76.A65T43   1994
005.3'028'7—dc20   94-21431
                                                           CIP

Printed in the United States of America
10 9 8 7 6 5 4 3 2 1

Contents

152, 112

Preface

The term *technology* conveys the idea of hardware, the silicon and plastic computer platforms. Our usage is more general, technology defined as a "systematic treatment" or "applied science." For a process to be labeled a technology, it must be replicable, or able to be used repeatedly with the same consequences (Lumsdaine, 1963). We count as technology replicable procedures, with or without hardware. For example, empirically validated procedures used to organize teams of students for learning tasks are a form of technology, even though no hardware is required. Labeled soft technology, any computer software would fit this definition of technology.

This volume represents an expansion of ideas presented at the conference "Technology Assessment: Estimating the Future" held at the University of California, Los Angeles. The goal for the conference was to identify the major strategies and more promising practices for assessing technology. Perspectives were shared by leaders in computer science, cognitive and military psychology, and education. The speakers, representing government, business, and university sectors, helped to set the boundaries of present technology assessment. The conference proved to be very much work-in-progress in its focus and underscored the relatively little systematic thought given to the assessment (and evaluation) of technologies of all sorts. A subset of the conference is provided in this book. The chapters were updated, situated in the context of existing research, and extensively reviewed and edited. Two volumes resulted—this one concerning technology assessments of software systems, and another dealing with assessments in education and training (see Baker & O'Neil, 1994).

Technology Assessment is designed for professionals in the assessment community, broadly defined to include psychology, computer science, and engineering disciplines. It explores both the use of techniques to assess technology and the use of technology to facilitate the assessment process.

In summary, the purpose of this book is to portray the state of the art in technology assessment and to provide conceptual options to help us understand the power of technology. Technological innovation will continue to develop its own standards of practice and effectiveness. To the extent that these are empirically based, designers, supporters, and consumers will be given better information for their decisions.

ACKNOWLEDGMENTS

This book could not have come into existence without the help and encouragement of many people. Our thanks to our editor, Hollis Heimbouch of Lawrence Erlbaum Associates, Inc., for her support and guidance in the publication process. We thank Katharine Fry for her magical assistance in preparing the manuscript.

This research was supported in part by a contract (N00014-86-K-0395) from the Defense Advanced Research Projects Agency (DARPA), administered by the Office of Naval Research (ONR), to the UCLA Center for the Study of Evaluation/Center for Technology Assessment. However, the findings and opinions expressed do not necessarily reflect the positions of DARPA or ONR, and no official endorsement by these two organizations should be inferred. This research also was supported in part under the Educational Research and Development Center Program cooperative agreement R117G10027 and CFDA catalog number 84.117G as administered by the Office of Educational Research and Improvement, U.S. Department of Education. The findings and opinions expressed in this report do not reflect the position or policies of the Office of Educational Research and Improvement or the U.S. Department of Education.

REFERENCES

Baker, E. L., & O'Neil, H. F., Jr. (Eds.). (1994). *Technology assessment in education and training.* Hillsdale, NJ: Lawrence Erlbaum Associates.

Lumsdaine, A. (1963). Instruments and media of instruction. In N. L. Gage (Ed.), *Handbook of research on teaching* (pp. 583–682). Chicago: Rand McNally.

Introduction

Harold F. O'Neil, Jr.
University of Southern California/CRESST

Eva L. Baker
CRESST/University of California, Los Angeles

The impact of technology on all aspects of contemporary life is an unchallenged fact. We alternately suffer with and revel in the side effects of technology: from toxic waste, the social effects of too much television, and carpal tunnel syndrome, to the wonders of spreadsheets, tiny video cameras, and *WD-40*. Yet technology effects, both good and bad, seem always to take us by surprise. Why don't we systematically look at technology and better anticipate its consequences? In fact, many scholars do study technology, but most of our methods are weak and our attention at best happenstance. Limited in many ways by a gee-whiz view of technology, we greet every new "advance" as unalloyed good. It is only later we feel we may have been tricked. Paradoxically, the task of technology development itself affects the assessment of its utility. Technology serves to codify and automate procedures and to extend our sphere of action. Unlike scientific findings, which are subject to interpretation, we act as if technology development is binary: either it works or it doesn't, it runs or it crashes, it delivers or fails. Thus for many designers and developers, technological innovation is an existence proof. Look at what we made! And the novelty either does or doesn't capture our imagination.

Assessing a technology involves making an estimate or judgment of its current state with regard to its effectiveness in a foreseeable range of applications. Technology assessment involves, therefore, at least two levels of prediction: first, predicting from the current status of technology to future utilities (e.g., U.S. Department of Commerce, 1990); second, generalizing from a specific case or cases to unknown applications to be invented in the future. These leaps in generalizability typically do not worry experimentalists, who engage in their predictions using conventions of science. Scientific safeguards include the cre-

1

ation of sampling distributions of populations to which results will generalize, and the conception of what is studied as variables rather than as objects. In the case of assessment of individual technologies, these safeguards are not in place. Thus, it is entirely probable that an evaluation of a single technological enterprise, such as a computer program, might be misleading in the prediction of its utility to future contexts or to other cases or instances of application. On the one hand, the particular case of technology might be rapidly superseded by another approach, vitiating its potential contribution. For example, the creation of video cameras destroyed any predictions one might have wished to make about hand-held motion picture cameras. On the other hand, the example of the technology application assessed may only bear limited relationship to its successive versions—for instance, in text editing software.

Thus, the methods available for technology assessment appear to be relatively weak. One prominent method focuses on demonstration. A developer creates a prototype or demonstration model of an innovation, for instance, a networking protocol. This protocol is demonstrated with mock or real objects and data to a potential set of sponsors or users. Based on their reactions, the technology is assessed in terms of its utility in meeting stated requirements, its identification and satisfaction of new requirements, or its ease and efficiency compared to an existing option. A stronger version of this demonstration involves actual operational versions of the technology. For example, the Department of Defense demonstrated the utility of its new iteration of distributed interactive simulation (Alluisi, 1991; Berry, 1991) by connecting up actual air, sea, and land resources with simulated resources for a nationally televised congressional hearing. Demonstrations are existence proofs. They translate concepts into a version of reality that may be especially convincing to novices.

A second class of technology assessment enterprises is based on explicit examples of the evaluation of technology. In such instances, an innovation, such as a new approach to teaching mathematics, or a user interface, is tested empirically. Beyond the ability to demonstrate its processes—that is, "to run"—the system or innovation is judged on the extent to which it meets its stated goals. Users are typically assessed, and their judgments of the ease and efficiency of operation may be used as data for the study (Chorafas, 1990).

To get an entire picture of a class of technology, rather than making inferences from piecemeal evaluations using a one-innovation-at-a-time approach, formal meta-analytical approaches may be used, where studies or empirical evaluations of individual technologies are statistically combined to give the status of the population (Kulik, 1994). Such studies permit one to understand the reliability or consistency of findings and to learn something about the generalizability of the results to settings or requirements differing in specific dimensions, such as age of user, degree of support, and so on.

Finally, there are assessment approaches that break away from standard evaluation techniques and try to find new metaphors for characterizing outcomes. As an example, we present a suite of studies that evolved from the assessment of artificial intelligence technology, sponsored by the Defense Advanced Research Projects Agency (DARPA) and the Office of Naval Research. The metaphor we used in those studies was human benchmarking. In this volume, the chapters by O'Neil, Baker, Ni, Jacoby, and Swigger (chap. 1); Skrzypek (chap. 2); and Baker (chap. 3) are examples of human benchmarking.

It may be relevant to provide some history of how this particular array of research in human benchmarking, called the artificial intelligence measurement system, developed and what problems one may anticipate in the design of technology assessments of high-end technology.

ARTIFICIAL INTELLIGENCE MEASUREMENT SYSTEM

The artificial intelligence measurement system (AIMS) project was undertaken as an exploration of methodology to investigate how the effects of artificial intelligence systems could be compared to human performance. It was designed under a number of assumptions. First, that human performance is infinitely richer than the relatively primitive computer systems so far designed. Although the principal measurement strategy proposed treating system performance as if it were a point in a distribution of human performance, there was no intention of equating conceptually computer systems and individual human performance. Prior research by Clancey (1988), for example, documented the fact that computer systems because of their consistency and dependence on a coherent view (one or more experts) could be compared to a set of humans working on problems in a particular domain. Rather, the exploratory goal of this project was to investigate whether intelligent systems could be placed on a continuum of human performance.

In practice, this mapping would test some a priori correspondences, in that somewhat unsophisticated systems would be mapped on a sample of individuals with relatively low performance and more sophisticated systems would map to individuals with more sophisticated levels of performance. If such a set of rough correspondences could be established, then it would be theoretically possible to benchmark systems under development in terms of progressively higher performing populations of individuals. Effectiveness, in terms of a performance and investment ratio, could be judged for increasingly expensive implications. As a simple example, we could imagine comparing the mathematics problems solved by a system with the performance of students in kindergarten, sixth grade, and beginning calculus.

Originally, the project was formulated to focus in one area—natural language understanding with the corresponding human performance domain of reading comprehension. This area held much promise because of (a) the rich research in both natural language understanding and reading comprehension, and (b) the clear differentiation of individuals in terms of dimensions underlying text understanding. However, we were encouraged by our sponsors to consider multiple areas simultaneously: natural language understanding, including interfaces and texts; expert system shells and expert systems; and machine vision. The project also included a technology assessment component to permit reflection on our processes in the light of progress made elsewhere. This book is the result of such reflection.

There was also the assumption that the project would in part depend on collaboration with members of the computer science discipline. It was also assumed that this requirement would provide a challenge because the form of evaluation we were exploring would not be within the expectations or values of members of this discipline. Although we experienced difficulties in acquiring systems for use and in sustaining interest of some computer scientists, critical components of this work were led or strongly influenced by members of the computer science community. Moreover, the project had a desired effect in energizing members of the community to explore approaches beyond standard software metrics to evaluate the impact of their efforts.

The project experienced all the usual difficulties in dealing with complex software—delays in hardware implementations, concerns about the proprietary nature of code—as well as some unanticipated problems, such as the requirement but inability to evaluate systems implemented in classified domains. Staff also needed to quell occasional anxiety attacks related to imagined litigation occasioned by the public evaluation of commercial products.

As a strategy, the project invested the bulk of its resources in the natural language area. There it focused on two different types of implementation: interfaces that served to query data bases (see Baker, chap. 3) or as front-ends to expert systems, and experimental text understanding systems (Butler et al., 1990). A principal effort in this project component was the development of a compatible descriptive/empirical strategy. The creation of a sourcebook of problems in natural language (Read et al., 1990) was undertaken as a way to describe and map the field. This sourcebook could provide an interpretative context for the understanding of any empirical benchmarking results in natural language. Thus, the empirical benchmarking of natural language systems could be understood in terms of the difficulty of the task. A partial analogy is the degree of difficulty score paired with the performance score for a diver. Description was also a key element in the other project components as well, although nowhere was the effort as extensive as in the natural language understanding tasks. The machine vision project also created a sourcebook of problems (Skrzypek, Mesrobian, & Gungner, 1988b) and described existing vision systems and measures (Skrzypek, chap. 2). The expert system project created a framework for both expert systems and analogous human processes (O'Neil et al., chap. 1).

The empirical human benchmarking strategy was originally predicated on the idea that existing, commercially available, or research-validated achievement tests would be used to benchmark (or compare) multiple implementations. Early in the project, it became clear that, except in the area of vision, no available tests reflected the domains used in particular implementations. Although technical strategies such as linking and equating can be used to combine information from disparate tests, these techniques impose constraints in terms of the underlying dimension(s) to be measured and require large sample sizes (Petersen, Kolen, & Hoover, 1989). A commercial test was used to assess reading ability, but the project required unanticipated effort in test development to create a comparable performance base for comparison. The development of new measures followed procedures identified in Hively, Patterson, and Page (1968) and in Baker and Herman (1983), producing what is known as domain-referenced achievement tests.

In the natural language area, an additional attempt was made to overcome the domain specificity problem. In our benchmarking of a front-end query system, we created a measure that dissociated the structure of the query from its content base (Baker, chap. 1). Focusing on type of query allowed us to measure students' understanding of the linguistic structure rather than the particular content domain in which the front-end system was implemented, a classified domain of navy information. In other test development areas, we were able to sidestep the domain issue by focusing on process, for example, the development of a test of metacognitive strategy described in the expert system component (O'Neil et al., chap. 1). However, for much of our effort we were very much focused on the domain of task, on the particular texts in systems, or on the particular content area of an expert system.

NATURAL LANGUAGE UNDERSTANDING

In more detail, our research in the area of natural language understanding focused on methods of evaluating natural language processing (NLP) systems, with a twofold goal:

1. We were interested in the identification and classification by example of problems in natural language understanding.
2. We were interested in the development of an evaluation methodology that considers system output relative to or benchmarked to human performance.

The first approach took into account the processes that lead to output; the second approach was concerned with output only. These two evaluation metrics can be used to describe natural language processing systems in complementary ways.

The first approach to the issue of natural language processing system evaluation, that of identification by example and classification of problems in natural language understanding, is realized in practical form in the *Natural Language Sourcebook* (Read et al., 1990). This is a collection of 197 examples of natural language processing problems organized by a classification scheme that reflects an artificial intelligence perspective, and cross-referenced by two other classification schemes, one reflecting a linguistic perspective and the other a cognitive-psychological perspective on the types of issues presented in the examples.

The sourcebook developmental process involved a search through the artificial intelligence, computational linguistics, and cognitive science literature to identify examples of processing problems. Each example served as the basis for a sourcebook entry. The entries, called "exemplars," each consist of (a) one or more sentences, a fragment of dialogue, or a piece of text that illustrates a conceptual issue; (b) a reference; and (c) a discussion of the problem a system might have in understanding the example. An example is used to illustrate each problem, followed by a discussion that defines the type of problem by delineating the information-processing issues involved. The sourcebook exemplars provide discussions of concrete processing problems in terms of the general principles. This grounding of the general in the specific makes the sourcebook a uniquely useful and appropriate tool for evaluation of natural language processing systems. Once the exemplars were completed, a linguistic and cognitive-psychological cross-index was added.

BENCHMARKING TO HUMAN PERFORMANCE

The second approach to the issue of natural language processing system evaluation, that of evaluating natural language processing systems by benchmarking them to human performance, was explored in two major studies. The first study provides an initial specification of a continuum of difficulty for language that a syntactic shell interface, IRUS, can process (Baker, chap. 3; Baker, Turner, & Butler, 1990). The continuum of difficulty is based on the performance of kindergartners and first-graders on comprehension tasks syntactically parallel to those accomplished by IRUS and is reported in this volume.

The second study provides a comparison of the abilities of six text understanding systems to answer specific questions about given texts with the abilities of humans to answer the same questions about the same texts (Butler et al., 1990). In this study, systems were benchmarked to grade-equivalent groups of human subjects. In Baker et al. (1990), correct responses for the human subjects were determined by how IRUS responded to parallel items (i.e., all the IRUS responses were taken to be correct), whereas in Butler et al. (1990), correct responses for both human subjects and intelligent computer systems were determined by the consensus responses of adult native speakers.

EXPERT SYSTEM SHELLS

A second major component of the project investigated reasonable approaches to the evaluation of expert system shells. It explored (a) available social science methodologies that might be brought to bear on the study of expert system shells, and (b) the feasibility of implementing these strategies in a routine way.

This project began with the analysis of costs and benefits of experimental approaches to the study of expert systems, particularly the construction of a complex experiment in which shells and tasks were manipulated and randomly assigned to system developers with various levels of expertise. However, even supposing that critical variables such as expertise, order, domain knowledge, and task generalizability could have been controlled, we rejected the initial approach because of feasibility concerns—time, cost, and the small likelihood that system developers appropriate to represent the population of interest could be released from their regular tasks in order to complete our experimental requirements.

Instead, we decided to take a different tack and assess qualitatively the process of knowledge engineering and system development using a case study approach. Following a review of the literature (see Novak, Baker, & Slawson, 1991), the project recognized that typical software metrics in use for shell evaluation (e.g., Richer, 1985) did not focus in detail on the processes or the outcomes of development. Although our literature review did turn up studies focused on user satisfaction and consumer-guide sorts of analyses, in-depth studies of knowledge engineering processes had not been made. Consequently, the project posited the idea of developing a 2×2 design for the conduct of intensive case studies, with one factor focusing on the sophistication of the shell in terms of representation and inferencing strategies and the other factor focusing on the nature of the problem, whether it was well defined or ill-structured. To undertake this work, a well-defined problem, selecting the appropriate reliability index for use with a particular form of achievement test, was formulated. An expert psychometrician was identified and videotapes and observations of the knowledge engineering process were made. The first system employed was relatively unsophisticated, M-1™ (Li, 1987, 1988; Slawson, Novak, & Hambleton, 1988). The prototype implementation was reviewed by the expert and found to be unsatisfactory because of domain misconceptions by the knowledge engineer. Rather than proceed to completion, the expert recommended that we try something else. Using the existing videotapes and with minimal visits with the expert, an implementation was made by another programmer of an expert system using NEXPERT™ (Novak et al., 1991). At that point, given the difficulty and cost of this strategy, with the approval of our advisors, we stopped research on shells, instead focusing on expert systems.

BENCHMARKING EXPERT SYSTEMS

The problem of human benchmarking in an expert system context was addressed by research attending to the following questions:

1. What descriptive analyses of computer expert processes and human cognitive processes should be attempted?
2. On what dimensions could expert system performance be benchmarked on humans?

The project began with a literature review of benchmarking of expert systems (O'Neil, Ni, & Jacoby, 1990) in which it became clear that the project could opt to have computer-science-driven models or psychologically driven models of benchmarking. Although it would be ideal to cross-validate these approaches, we were constrained by the low number of expert system implementations that permit multiple tests of a psychologically driven measurement model. The decision was to conduct human benchmarking according to the conceptual model originally outlined in Baker (1987; chap. 3, this vol.), that is, to norm an expert system's performance on samples of individuals.

Expert systems always involve considerable amounts of domain-specific knowledge; thus, unlike the IRUS work described earlier, it was difficult to isolate the structure of tasks from their content. Through the use of metaphor, we transformed the essence of an expert system (GATES, a system that assigned airplanes to gates in a major airline hub) into a valid psychological construct. The GATES program schedules by assigning an airplane to a time, location, and so forth, without violating constraints. The psychological equivalent of this task is called self-monitoring in the psychological literature. We surveyed the extant measurement literature to identify an existing, high quality instrument to assess this aspect of human metacognition. Finding none, we developed one ourselves. A study was designed that incorporated both the benchmarking of outcomes (how well samples of students completed the GATES tasks) and of human processes (how well students planned, selected strategies, and monitored their behavior while executing the task, and how aware they were of these processes) (O'Neil, Ni, Jacoby, & Swigger, 1990). Finally, a report of the evaluation, using both process and outcome measures, was prepared following the conduct of experimental trials (O'Neil et al., chap. 1). The methodology was demonstrated to be successful in that individuals with a priori different ability levels performed predictably. A summary of the entire set of activities is included in O'Neil, Baker, Jacoby, Ni, and Wittrock (1990).

MACHINE VISION

The machine vision benchmarking component was completed under the direction of Dr. Josef Skrzypek of the UCLA Computer Science Department. This component sought to answer the following questions:

1. As a long-term goal, how might the benchmarking of machine vision proceed as a joint effort between the neurosciences and computer science?
2. Specifically related to this project, can a framework be generated for evaluating progress in machine vision by documenting the status of the field and investigating the human visual performances that could be benchmarked on a vision system?

The strategy used for the vision benchmarking component, initially described in Baker (1987) and in Baker, Lindheim, and Skrzypek (1988), in some ways paralleled the strategy used in the natural language component (Baker, chap. 3). The project conducted an extensive review of 15 vision systems in order to identify possible categories along which machine vision systems could be evaluated. In the report by Skrzypek, Mesrobian, and Gungner (1988a), each of these analyses is followed by justifications for the use of the human visual system as a model for a general purpose vision system. The report identifies visual tasks from existing tests and discusses them in terms of their corresponding computational neural substrates.

Comparisons among systems are made along five dimensions: (a) image attributes, (b) perceptual primitives, (c) knowledge base, (d) object representation, and (e) control. Skrzypek and his colleagues rejected the attempt to benchmark individual vision systems directly for a number of reasons. One constraint was the idiosyncratic platforms used in the development of such systems. The cost of acquiring such sufficient hardware appropriately configured was well beyond the resources of their project. Similarly, the particular domain of interest for these systems was extremely narrow. When approaching the problem from the human side, benchmarking ran into some limitations, in large measure because the bulk of existing systems focused on lower and middle range visual tasks with minimal cognitive demands. Such tasks were outside accessible ranges for typical individuals. Simple tasks, like matching to samples used in manufacturing systems, are tasks completed by people automatically and without awareness. The research team created a model of a general purpose vision system. They assembled typical visual tasks provided to individuals in regular psychological tests, such as paper folding and block tests, and documented evidence in neuroscience connected to those tasks. Finally, they created a sourcebook (Skrzypek et al., 1988b) documenting data-level visual tasks. Each entry consists of a problem statement, a discussion, references from the literature, and examples. Skrzypek (chap. 2) documents the work in machine vision.

ARTIFICIAL INTELLIGENCE MEASUREMENT
SYSTEM SUMMARY

The artificial intelligence measurement system project asked whether human benchmarking of computer systems is possible in a variety of classes of systems. Our answer to that question is yes. A corollary question is whether benchmarking processes are routinely feasible as evaluation procedures for intelligent systems. At the present time, our answer is no, for the practical and technical reasons alluded to earlier. We recommend the creation of descriptive resources, such as the *Natural Language Sourcebook* (Read et al., 1990), to enable the field to inform itself and to provide a mechanism to keep abreast of the progress made by the community. We recommend the further consideration of benchmarking when there are sufficient implementations in the same topic to support the investment in their common evaluation. Such evaluation would differentiate between emphases and effects of such systems by goal. Language accessible to research and development program managers and policymakers would be used, and benchmarks would be provided in terms of classes of human performance.

THE REMAINDER OF THE BOOK

A final component of the artificial intelligence measurement system effort was the attempt to be reflective and self-conscious about the strategies we undertook to evaluate complex education and training systems and software systems. These assessment strategies involve technical, social, financial, and policy dimensions. Clearly, working on the boundaries among the fields of computer science, military training, education, evaluation, and psychometrics provides a continuing challenge. The remaining chapters in this book provide different views of various technology assessments of software systems. A thumbnail description of each chapter serves as a map for readers of different interests.

Moore (chap. 4) reports the development and assessment of an explanation system. Her work is very much in the mode of an advanced demonstration of a concept. Swigger (chap. 6) provides a review of software assessment from a software engineering viewpoint. Seidel and Perez (chap. 7) provide a conceptual analysis of impact assessment based on numerous examples of technology development in the military. Their work is intended to affect the way assessment occurs over a range of specific interventions. Madni and Freedy (chap. 5) describe a system that changes the model by which engineering design occurs from a sequential one to a concurrent one. Feurzeig (chap. 8) covers visualization tools and also includes the description of an implementation, with a window on the logic used in its creation. Burns (chap. 9) takes an existing application, local area networks, and provides a rich description and set of examples of its extension to a new setting and use, in the service of technology assessment. Finally, Fletcher (chap. 10) provides an excellent example of a technology assessment of network simulation and its applicability to the assessment of teams.

ACKNOWLEDGMENTS

This research was supported in part by contract number N00014-86-K-0395 from the Defense Advanced Research Projects Agency (DARPA), administered by the Office of Naval Research (ONR), to the UCLA Center for the Study of Evaluation/Center for Technology Assessment. However, the findings and opinions expressed do not necessarity reflect the positions of DARPA or ONR, and no official endorsement by these two organization should be inferred. This research also was supported in part under the Educational Research and Development Center Program cooperative agreement R117G10027 and CFDA catalog number 84.117G as administered by the Office of Educational Research and Improvement, U.S. Department of Education. The findings and opinions expressed in this report do not reflect the position or policies of the Office of Educational Research and Improvement of the U.S. Department of Education.

REFERENCES

Alluisi, E. A. (1991). The development of technology for collective training: SIMNET, a case history. *Human Factors, 33*(3), 343–362.

Baker, E. L. (1987, March). *Artificial intelligence measurement system* (briefing charts). Presentation at the ONR Contractors' Meeting, Princeton, NJ, Yale University.

Baker, E. L., & Herman, J. (1983). Task structure design: Beyond linkage. *Journal of Educational Measurement, 20*(2), 149–164.

Baker, E. L., Lindheim, E. L., & Skrzypek, J. (1988, May). *Directly comparing computer and human performance in language understanding and visual reasoning* (CSE Tech. Rep. No. 288). Los Angeles: University of California, Center for Technology Assessment/Center for the Study of Evaluation, and Artificial Intelligence Laboratory, Computer Science Department.

Baker, E. L., Turner, J. L., & Butler, F. A. (1990). *An initial inquiry into the use of human performance to evaluate artificial intelligence systems*. Los Angeles: University of California, Center for Technology Assessment/Center for the Study of Evaluation.

Berry, C., Jr. (1991). Recreating history. The battle of 73 Easting (pp. 3–9). *National Defense* [Journal of the American Defense Preparedness Association], November.

Butler, F. A., Baker, E. L., Falk, T., Herl, H., Jang, Y., & Mutch, P. (1990). *Benchmarking text understanding systems to human performance: An exploration*. Los Angeles: University of California, Center for Technology Assessment/Center for the Study of Evaluation.

Chorafas, D. N. (1990). *Knowledge engineering*. New York: Van Nostrand Reinhold.

Clancey, W. J. (1988). Acquiring, representing, and evaluating a competence model of diagnostic strategy. In M.T.H. Chi, R. Glaser, & M. J. Farr (Eds.), *The nature of expertise* (pp. 343–418). Hillsdale, NJ: Lawrence Erlbaum Associates.

Hively, W., Patterson, H. L., & Page, S. A. (1968). A "universe-defined" system of achievement tests. *Journal of Educational Measurement, 5,* 275–290.

Kulik, J. A. (1994). Meta-analytic studies of findings on computer-based instruction. In E. L. Baker & H. F. O'Neil, Jr. (Eds.), *Technology assessment in education and training*. Hillsdale, NJ: Lawrence Erlbaum Associates.

Li, Z. (1987). *An expert system for selecting the index of reliability.* Los Angeles: University of Southern California, School of Education, Department of Instructional Psychology and Technology.

Li, Z. (1988). *Knowledge engineering report: An expert system for selecting reliability index.* Los Angeles: University of Southern California, School of Education, Department of Instructional Psychology and Technology.

Novak, J. R., Baker, E. L., & Slawson, D. A. (1991). *The evaluation of expert system shells.* Los Angeles: University of California, Center for Technology Assessment/Center for the Study of Evaluation.

O'Neil, H. F., Jr., Baker, E. L., Jacoby, A., Ni, Y., & Wittrock, M. (1990). *Human benchmarking studies of expert systems.* Los Angeles: University of Southern California, Cognitive Science Laboratory and University of California, Center for Technology Assessment/Center for the Study of Evaluation.

O'Neil, H. F., Jr., Ni, Y., & Jacoby, A. (1990). *Literature review: Human benchmarking of expert systems.* Los Angeles: University of Southern California, Cognitive Science Laboratory and University of California, Center for Technology Assessment/Center for the Study of Evaluation.

O'Neil, H. F., Jr., Ni, Y., Jacoby, A., & Swigger, K. M. (1990). *Human benchmarking methodology for expert systems.* Los Angeles: University of Southern California, Cognitive Science Laboratory; University of California, Center for Technology Assessment/Center for the Study of Evaluation.

Petersen, N. S., Kolen, M. J., & Hoover, H. D. (1989). Scaling, norming, and equating. In R. L. Linn (Ed.), *Educational measurement* (3rd ed., pp. 221–262). New York: Macmillan.

Read, W., Dyer, M., Baker, E., Mutch, P., Butler, F., Quilici, A., & Reeves, J. (1990). *Natural language sourcebook.* Los Angeles: University of California, Center for Technology Assessment/Center for the Study of Evaluation and Computer Science Department.

Richer, M. H. (1985). *Evaluating the existing tools for developing knowledge-based systems* (Rep. No. KSL 85-19). Stanford, CA: Stanford University, Stanford Knowledge Systems Laboratory.

Skrzypek, J., Mesrobian, E., & Gungner, D. (1988a). *Defining general purpose machine vision: Metrics for evaluation.* Los Angeles: University of California, Computer Science Department.

Skrzypek, J., Mesrobian, E., & Gungner, D. (1988b). *Machine Perception Laboratory visual task sourcebook.* Los Angeles: University of California, Computer Science Department.

Slawson, D. A., Novak, J., & Hambleton, R. K. (1988, April). *A qualitative approach to the evaluation of expert system shells.* Paper presented at the annual meeting of the American Educational Research Association, New Orleans.

U.S. Department of Commerce. (1990). *Emerging technologies. A survey of technical and economic opportunities.* Washington, DC: U.S. Department of Commerce, Technology Administration.

Slawson, D. A., Novak, J., & Hambleton, R. K. (1988, April). *A qualitative approach to the evaluation of expert system shells.* Paper presented at the annual meeting of the American Educational Research Association, New Orleans.

U.S. Department of Commerce. (1990). *Emerging technologies. A survey of technical and economic opportunities.* Washington, DC: U.S. Department of Commerce, Technology Administration.

1 Human Benchmarking for the Evaluation of Expert Systems

Harold F. O'Neil, Jr.
University of Southern California/CRESST

Eva L. Baker
Yujing Ni
Anat Jacoby
CRESST/University of California, Los Angeles

Kathleen M. Swigger
University of North Texas

BACKGROUND

The measurement of expert systems development efforts and systems challenges existing technologies for software assessment. We view an expert system as an artificial intelligence-based program that performs a task that a human expert can accomplish. The challenges for assessment stem from the knowledge engineering approach to software development and the nature of the expert system applications typically developed. Standard approaches to verification, validation, and quality assurance do not adequately address these challenges (Geissman & Schultz, 1988; Swigger, chap. 6).

Two kinds of problems result from the novel characteristics of expert system development. The first kind of problem stems from differences in the software engineering process. For example, existing approaches to software evaluation in the military are based on a model of conventional software development. But the development of knowledge-based expert systems may not conform to the phases in the development life cycle as described in DOD-STD-2167A (U.S. Department of Defense, 1987). For example, expert system development is more distinctly incremental. Specifications are often developed in the course of knowledge engineering. Thus the verification approach of assessing the simple match between preordained specifications and code becomes inappropriate. Even if a standard verification and validation approach could be taken, the development phases and milestones for expert systems development do not align with those used in conventional software engineering.

One way in which the process differences may lead to differences in measurement methodologies is that expert system development does not lend itself well to the use of written formal specifications. Iterative design and coding tends to be more of a data-driven, as opposed to requirements-driven, enterprise. Many details of functional specification emerge only in the course of development. Although many software attributes such as speed requirements or user interface design may be specified in advance, the contents of a knowledge base and the implicit inferencing logic—that is, the program itself—are the only complete statement of the software's problem-solving capabilities. This absence of specifications means that assessment cannot simply be comparisons of specifications and outputs, or specifications and code. Rule explanations, whether explicitly elicited and coded or generated by the shell (e.g., via inference trees), constitute the documentation for the origin and purpose of rules. The effect of these rules under different inputs and inference methods determines the problem-solving behavior of the system. Thus the task of evaluation of problem-solving functionality becomes one of "reverse engineering" the performance specifications from the functioning knowledge base.

The second kind of problem in expert system evaluation is that there is no agreed upon or fully developed methodology for assessing expert system attributes and quality (U.S. Department of Defense, 1987). This deficiency can, in part, be attributed to the relative newness of the development approach, as already suggested. But characteristics of the products of these development efforts also pose special problems. Because the output, if not the process that is still hard to model (Kopcso, Pipino, & Rybolt, 1988; Tonn & Goeltz, 1990), of an expert system is intended to replicate human expert performance, the problem of defining and measuring expertise arises. Performance criteria or standards for expert systems are difficult to define. Hayes-Roth, Waterman, and Lenat (1983) noted that the assessment of human expertise is just as troublesome. They cited the need for a "gold standard," or performance criterion, against which expert systems can be judged. This need is still present.

A measurement model for expert systems not only will need to account for alternative user contexts, decisions, resources, and goals, but also may serve three functions:

1. Provide approaches to evaluate candidate tools (e.g., "shells") for those frequent occasions when expert system development is assisted by commercial products.
2. Provide approaches to evaluate the process of development at each stage.
3. Provide approaches to evaluate the quality of the product and process of expert systems.

Any measurement model needs to account for a full range of factors that influence the utility of the resulting information. Among the most salient is the type(s) of decisions of interest. At issue is whether the intent of the evaluation is summative (that is, to make a choice among competing products, as in Functions 1 and 3) or formative (that is, to influence the quality of the process and product as it is developed). A measurement model should also encompass both quantitative and qualitative data collection and analysis techniques. In addition, a measurement model should provide options that are low cost as well as those that are resource intensive. The best information can usually be obtained at a high cost. It is critical that a model optimize cost of the measurement process with risk and criticality of the expert system application; for example, exploratory systems may only need low-cost techniques. This chapter focuses on Function 3.

Our human benchmarking approach is to conduct an evaluation, that is, to norm an expert system's performance on a sample of people's performance. The importance of this approach is that it goes beyond the conventional approach of expert system evaluation and aims to build psychometric criteria through comparison of an expert system's performance with differentiated performances by different samples of people.

Benchmarking is defined by Guralnik (1984) in the following manner: "(1) Surveyor's mark made on a permanent landmark of known position and altitude, used as a reference point in determining other altitudes, (2) standard or point of reference in measuring or judging quality" (p. 131). Another definition of benchmarking is provided by Camp (1989): "The approach of establishing operating targets and productivity programs based on industry best practices leads to superior performance. That process, being used increasingly in U.S. business, is known as *benchmarking*" (p. 3). Benchmarking is also used to denote a standard process for measuring the performance capabilities of software and hardware systems from various vendors (Benwell, 1975; Letmanyi, 1984). It is in the latter sense that we use benchmarking; thus, human benchmarking is an evaluation procedure by which an expert system's performance is judged based on a sample of people's performance on tasks with psychological fidelity.

There are two major human benchmarking alternatives: computer science driven or psychological process driven. The computer-science-driven approach is either (a) expert system driven, in which one picks an expert system that encodes a human expert, derives psychological processes, and tests the processes with people; or (b) expert system shell driven, in which one estimates the "intelligence" of the shell (parent), assumes that resulting expert systems will have similar "intelligence," then follows the procedures of the expert-system-driven approach.

The second alternative for human benchmarking is psychological process driven in that one picks a psychological process, finds an expert system that exemplifies the process, and then tailors a test for the expert system and a test using people. For us, the psychological process approach was not feasible due to a constraint of available expert developers with robust, documented programs who were willing to collaborate with us. Thus, we chose the computer-science-driven approach. In turn, because our effort is focused on expert systems, we chose this subapproach rather than focusing on shells.

In order to provide an intellectual foundation for human benchmarking, a literature review was conducted and documented (O'Neil, Ni, & Jacoby, 1990). The literature was reviewed from two viewpoints: (a) a computer science and software engineering perspective and (b) a cognitive science perspective with a focus on psychological assessment.

This literature review suggested there are different approaches to expert system evaluation, including evaluation criteria and evaluation procedures. The literature offered diverse environments to capture the developmental aspects of expert systems evaluation. The review suggested the possibility of developing a psychometric standard for the evaluation of expert systems; it also helped us to document similarities and differences between cognitive psychology and artificial intelligence, which is important for our human benchmarking approach. Further, it suggested multiple frameworks useful for assessment.

Our basic idea of a human benchmarking approach was to use people's performance(s) to evaluate an artificial intelligence system. On this line of thought, the literature suggested the following three methods for the approach: (a) the Turing test, (b) comparison of the system's performance with performance of persons with different expertise in the same field, and (c) evaluation of the system's performance against a norm generated from the assessment of performance by groups from both inside and outside the relevant field. These methods differ in their samples of people and their evaluation criteria. The sample of people in the Turing test is only the expert group; in the second method, the sample includes people with different levels of expertise in the same field. In the third method, both experts and nonexperts comprise the sample from both inside and outside the relevant field. In terms of evaluation criteria, the first two methods usually use prespecifications for a system as evaluation criteria, whereas the third uses psychometric-oriented criteria generated from test specifications based on a structural analysis of a domain task. Due to the features in the sample of people and the evaluation criteria, the third method appears to have a potential generalizability that may lead to directly evaluating a system's "intelligence" and carrying out comparisons among similar or even different kinds of systems. It is this methodology that we attempted to explore, which is reported in the following three parts: evaluation of expert systems, modification of human benchmarking methodology of expert systems, and expert system application of human benchmarking.

To provide a context for our human benchmarking approach, the first part provides a literature review of the current status of expert system evaluation. The second part describes an initial study of human benchmarking approach to a natural language understanding system and a process of modifying this method to extend to the evaluation of expert systems. The third part reports results of a pilot study and a formal study using a group of people's performance to evaluate an expert scheduling system, GATES.

EVALUATION OF EXPERT SYSTEMS

To investigate the current state of the art in evaluation of expert systems, we located a large number of studies that either dealt with expert system evaluation or included evaluation as one component of expert system development. We began the review process by computer searching four data bases, that is, Education Resource Information Center (ERIC), National Technical Information Service (NTIS), Applied Science and Technology (AST), and University Microfilm Abstracts (UMI) of dissertation abstracts. The searches yielded a set of relevant studies, including evaluation methods, empirical research, reviews of literature, and theoretical papers. One of the most interesting descriptions of evaluation methods was provided by Hollnagel (1989).

Evaluation Methods

Hollnagel (1989) summarized five categories of evaluation methods for expert systems and their respective advantages. These categories were the Turing test, expert assessment, statistical sampling, summative and formative evaluation, and an analytic hierarchy process.

Turing Test.   This is the classical way of evaluating expert systems. An expert, as a judge, compares the system's performance and human experts' performance without knowing the subject performers' identity. The MYCIN system was evaluated with this method (Yu et al., 1984).

Expert Assessment.   This is criterion-referenced assessment. An expert system is assessed by an expert judge to compare the system's performance with clearly predefined criteria (either an expert's performance or a historically set standard) in an absolute way rather than based on a comparison with another system or expert's performance (Hudson & Cohen, 1984).

Statistical Sampling.   This means statistical sampling of test cases. The test cases for the assessment of performance by an expert system are statistically sampled, so that the chosen cases are known to be representative (Hudson & Cohen, 1984; Yu et al., 1984).

Summative and Formative Evaluation.   Summative evaluation (Scriven, 1967) focuses on overall choices among systems or programs based on performance levels, time, and cost. This evaluation is essentially comparative and contrasts the innovation against other options. The typical questions the evaluation asks are "Does the intervention work?", "How much does it cost?", and "Should we buy it?" Formative evaluation (Baker, 1974) seeks to provide information that focuses on the improvement of the innovation and is designed to assist the developer. Formative evaluation also addresses, from a meta-evaluation perspective, the effectiveness of the development procedures used in order to predict whether the application of similar approaches will likely have effective and efficient results (O'Neil & Baker, 1987). Thus, the formative evaluation seeks to improve the technology at large, rather than the specific instances addressed one at a time.

Analytic Hierarchy Process.   The expert system is decomposed into constituent components to see each part's performance (Liebowitz, 1985).

Hollnagel (1989) then compared these methods by referring to the criteria in Table 1.1. These criteria are correctness of the reasoning techniques, sensitivity, robustness, correctness of the final decision, accuracy of the final decision, quality of the human–computer interaction, and cost-effectiveness. These methods focus on the evaluation of an expert system after its implementation. Our human benchmarking approach also evaluates an expert system after implementation.

The essential task of expert system evaluation is to translate selected criteria into testable requirements, determine variables or parameters constrained by the

TABLE 1.1
Components of System Evaluation

Method Evaluation	Evaluation Criteria
Reliability technique	Correctness of reasoning, or the internal consistency of the reasoning technique
	Sensitivity, or the minimum variation in input needed to change the outcome of the decision
	Robustness, or the ability to absorb and compensate for nonstandard input
Validity	Correctness of the final decision or output, consistent with the needs in the given contexts
	Accuracy of the final decision, or the extent to which the consequences of alternatives are satisfactory
Usability	Quality of the human-computer interface, or the degree to which the interaction between users and system functions effectively
	Cost-effectiveness, or the gain from the use of the system related to its cost

Note. Adapted from *Topics in Expert System Design: Methodologies and Tools* (p. 412) by E. Hollnagel, 1989, New York: Elsivier Science Publishers, North Holland. Copyright (1989) by Elsevier Science Publishing Company. Adapted by permission.

requirements, and design test cases for the manipulation of the variables or parameters. However, the most serious problem in carrying out these procedures is that testable requirements are hard to define because relatively few requirements are initially known but are formulated during or after the development of an expert system. Related to this problem is that the profile of representative test cases is usually unknown. Thus there is great uncertainty in whether a set of selected test cases represents a reasonable range of example collection.

It is worth mentioning that recently the need is stressed for considering expert system evaluation throughout the development process, not only after implementation. Furthermore, a social–technical framework, which treats a working system as a combination of the social and technical components, is proposed as an alternative to the restrictive verification and validation techniques to assess long-term effects of expert systems (Berry & Hart, 1990; Sharma & Conrath, 1992).

Empirical Studies in Evaluation of Expert Systems

In our selected set of references, there were few empirical studies. The methods they employed included the Turing test, expert assessment of face validation, criteria assessment with either known results or established test cases, field testing, subsystem validation, human benchmarking, summative and formative evaluation, and impact analysis (effects of expert system uses on people's performance and organizations' effectiveness and efficiency such as workload, accuracy, and confidence). The test cases in these studies either were generated by one or more experts in correspondence with their perceived prespecification for a system (e.g., Hushon, 1988), or were taken from previously used examples in evaluation of similar systems (e.g., Dillingham, Tam, & Kotras, 1988; Weiss & Kulilowski, 1988). It is also common for the domain expert to generate test cases to be evaluated during system development.

The establishment of a criterion-referenced testing technology for human performance assessment is a relatively recent phenomenon (Baker, O'Neil, & Linn, 1990) that should not be ignored. Test case generation today is often of the ad hoc variety, where the expert arbitrarily generates "typical" situations and solutions, and the knowledge engineer introduces variations in some parameters in order to test boundary conditions and other variations. The failure to circumscribe the domain in terms of performance criteria makes it difficult to ever know if all of the important variations have been tested. Criterion-referenced testing and item generation technologies used in human performance assessment could be of enormous benefit in designing test case domain specifications that align with applications requirements, specifying input and output conditions, and generating items that effectively sample the domains. Without some systematic approach to test case development, it is difficult to know the limits of an application or to be assured that gaping holes do not lurk within the code.

The representativeness of the cases seems to be problematic because the test-case generation technique relies only on the experts' arbitrary judgments, which vary across situations and over time. In fact, the representativeness issue occurs during the process of expert system development. The domain knowledge coded in an expert system is usually extracted from one or two experts in the field (e.g., the GATES system's domain knowledge is from an expert). However, there is great variation in domain experts for a given problem (e.g., Cochran & Hutchins, 1991); even the same expert may change his solution methods for the same problem over time. Thus, merely using prespecifications as evaluation criteria and a very small sample of test cases make the representativeness of the test cases uncertain.

Alternative approaches to evaluating expert systems involve the following: (a) the use of expert performance ratings; (b) use of test cases or benchmarks; and (c) tracking the effect on job performance, as was done on the MYCIN project (Shortliffe, 1976) and the XCON project (Sviokla, 1990). A number of expert raters or end users may make judgments about system performance, preferably in blind designs to avoid bias. When multiple raters are used, validation should include the assessment of interjudge consistency of ratings. For such metrics, generalizability theory, or "G-theory" (Brennan, 1983; Cronbach, Gleser, Nanda, & Rajaratnam, 1972), holds great promise. G-theory has been used successfully in military performance testing (Wigdor & Green, 1991) for analogous purposes (i.e., minimizing sources of variance with multiple raters and performance items) but has never been applied to expert system assessment.

Perhaps the ultimate test of an expert system is its effect on performance of the job it assists its user to accomplish. Assessment of expert systems via job performance can be measured using a controlled experimental method in a real or simulated task environment. Quality of decisions made on a diagnostic consultation task, for example, can be measured by expert ratings when the task is performed by an unaided novice, by users with various abilities aided by the software, or by experts alone. Significant differences with and without the software indicate its effectiveness or lack thereof. Significant interactions between software presence/absence and user characteristics would indicate important factors to consider for training and selection of users when the application is fielded.

MODIFICATION OF HUMAN BENCHMARKING
METHODOLOGY FOR EXPERT SYSTEMS

A human benchmarking approach was initially applied to a natural language understanding system (Baker & Lindheim, 1988; Baker, Turner, & Butler, 1990). To extend it to the evaluation of expert systems, we modified this method by relating cognitive psychological taxonomies to an expert system taxonomy (e.g., monitoring). With this bridge, we developed measures for groups of people to allow us to benchmark an expert system.

Natural Language Application of Human Benchmarking

The human benchmarking method was initially explored in the evaluation of a natural language system (Baker & Lindheim, 1988; Baker et al., 1990). The authors defined human benchmarking as assessing an artificial intelligence system in terms of similar performance by groups of people.

A natural language understanding system, IRUS, was selected as the target for the initial human benchmarking approach. IRUS is a human-computer interface that has natural language facility and allows a user to access a data base through a natural language by asking questions. A sample of IRUS queries was collected from a list of 163 queries used in an August 1986 navy demonstration of the capabilities of IRUS (Baker et al., 1990). A linguistic analysis of these queries served as the basis of test specifications for the development of the Natural Language Elementary Test (NLET). The Natural Language Elementary Test was designed to duplicate the queries that the system, IRUS, was able to understand. The sentence types IRUS was able to handle became the test specifications for the Natural Language Elementary Test. By the mediation of the test specifications, the Natural Language Elementary Test is "equivalent" to IRUS in linguistic functioning, for example, *How many cruisers are in WESTPAC?* (IRUS Query); *NP + Verb or Copula + Prepositional Phrase* (Linguistic Structure); *How many striped snakes are on the floor?* (Natural Language Elementary Test item).

The Natural Language Elementary Test has 39 items of similar linguistic structures to those of IRUS queries. The National Language Elementary Test was administered to two groups of students including kindergartners and first-graders. One way to benchmark IRUS at a particular level is to set an empirical criterion level, for example, 90% of the items, and set a performance standard so that 90% of the students would achieve that criterion. IRUS would be rated as first-grade level of natural language understanding (in a syntactic sense) if 90% of first-graders could understand all or most of Natural Language Elementary Test items. Because a sufficient number of levels could not feasibly be tested, a more practical approach to human benchmarking was a design to "norm" performance of the natural language system with differentiated performances by groups of people. Therefore, a standardized test was needed so that a Natural Language Elementary Test score could be interpreted or equated to a score of a standardized test. The score on the standardized test would have norms. Then, it would be possible to "norm" performance of natural language systems. Thus, the language section of the Iowa Tests of Basic Skills (ITBS), a standardized test, was also used in the natural language human benchmarking study.

It was expected that the first-graders would perform better than the kindergartners on both the Natural Language Elementary Test and the Iowa Tests of Basic Skills. This continuation then could be used as a "scale" to measure the performance of IRUS. However, the experimental results showed that these two groups' performance was similar on both tests. Thus, IRUS could be considered to function at the kindergarten or first-grade level of performance.

The natural language human benchmarking methodology relies heavily on domain-referenced testing as its central metaphor. The equivalence between IRUS queries and Natural Language Elementary Test items was determined in terms of common syntactic structures. Semantic equivalents were ignored.

However, one expects that human benchmarking of expert systems would be different than human benchmarking of natural language understanding systems. First, an expert system always involves a considerable amount of domain-specific knowledge; thus, unlike natural language systems, it is difficult in an expert system to isolate the structure of a task from its content for test specifications. For example, assigning airline flights to airport gates needs specific strategies as well as knowledge about the airport. In addition to capturing the requirements of domain-specific knowledge for a given task, one needs to take into consideration measurement of relevant experiences in human benchmarking of expert systems. For example, travel experience and scheduling experience would be relevant if one is to assign airline flights to airport gates.

Thus, as expected, we modified the methodology designed for a natural language application by augmenting it with metaphors and methodologies from the metacognitive skills literature. The basic idea was to map the methodologies and measures of human monitoring skills (e.g., Beyer, 1988; Weinstein & Mayer, 1986) onto expert systems in the area of scheduling. In order to achieve this goal, we required an instrument to measure human cognitive skills, something analogous to the Natural Language Elementary Test in the natural language area. Our preference was to select a reliable and valid commercially available instrument from the literature. Thus, our literature review (O'Neil et al., 1990) also focused on cognitive skills instruments.

Cognitive Skills Instruments

The search for cognitive skill instruments was intended to assist us in developing a human benchmarking approach to the evaluation of expert systems. The work began with an investigation of several cognitive and metacognitive skills as psychological equivalents of many of the process and outcome variables identified in the expert systems literature. Our selection reflected an integrative perspective from cognitive psychology and expert system applications. For example, "inference," "monitoring," "planning," "problem solving," and "reasoning" were considered because they are the same labels in expert system applications and cognitive psychology, whereas "diagnosis" and "scheduling" are from categories of expert system applications. Later, we concentrated effort in the areas of reasoning and metacognition. The main sources searched for these cognitive skill instruments were the Buros *Mental Measurements Yearbooks*

(1985–1989) and the Educational Testing Service (ETS) test catalogues (1986–1989). In addition, the Buros Mental Measurement Institute was contacted directly for information about unpublished tests.[1] Also, several active researchers (e.g., John Flavell and Richard Mayer) were contacted for their suggestions and advice. Finally, several relevant studies were located from the ERIC system. A 1992 update of this review, and the original review, revealed the findings discussed in the next subsections.

Monitoring.   There are two kinds of monitoring in cognitive activity. Monitoring, as a specific performance, refers to the cognitive activity of examining displayed status information, both formal (control panel displays) and informal (sounds, vibration, smells, etc.) (Moray, 1986; Parasuraman, 1986). As a general executive process, monitoring is directed at the acquisition of information about and the regulation of one's own, ongoing, problem-solving processes, which is described as "cognitive monitoring" by Flavell (1981), "metacognition" by Sternberg (1985), and "executive decision" by Kluwe (1987). These two levels of monitoring differ in that the former monitors the external world (e.g., a display) and the latter monitors the internal world (i.e., one's thoughts). Monitoring as an executive process can use information from monitoring one's thoughts of the external world (e.g., a display) or the internal world (e.g., one's thoughts). In this chapter, the term *monitoring* refers to the executive process of monitoring one's internal thoughts.

Unfortunately, we did not find a commercially available instrument for measurement of monitoring or metacognition. In some empirical studies, monitoring has been usually measured either through the subjects' self-reports of their cognitive processes during their performance or through some questionnaire about strategies they used in their learning and problem solving. We describe these approaches later.

Planning.   Planning, in cognitive psychology, is viewed as a problem-solving approach defined as the predetermination of a course of action aimed at achieving a goal (Hayes-Roth & Hayes-Roth, 1988); it is also considered as one component of metacognition regulating ongoing thinking processes (Beyer, 1988). It is also viewed as breaking a problem into a series of steps (Mayer & Wittrock, in press). We found no commercially available test for planning skills.

Diagnosis and Scheduling.   As mentioned earlier, diagnosis and scheduling are skills from the categories of expert system applications. There are no such categories in cognitive skills and thus no measures.

---

[1]The authors wish to thank Dr. Jane Conoley, Buros Institute of Mental Measurements, and Drs. John Flavell and Richard Mayer for their assistance.

TABLE 1.2
Comparison of AI and Psychological Constructs

AI	Psychology
Monitoring	
Monitoring observed behavior of a system to examine whether the function of a system is deviant from expected behavior (Waterman, 1986).	As a cognitive executive process directed at monitoring (e.g., self-checking one's own ongoing cognitive process) (Flavell, 1981; Kluwe, 1987; Sternberg, 1985).
Planning	
A problem-solving strategy that is defined as the predetermination of a course of action aimed at achieving a goal (B. Hayes-Roth & F. A. Hayes-Roth, 1979). Deciding on the entire course of action before acting, such as developing a plan for attacking enemy airfields (Waterman, 1986),	A problem-solving strategy that involves breaking a problem into a series of steps (Mayer & Wittrock, in press).
Diagnosis	
The process of fault finding in a system, based on interpretation of potential indicator data (F. Hayes-Roth, Waterman, & Lenat, 1983).	Not a specific research area.
Scheduling	
Selecting a sequence of operations needed to complete a plan, determining start and end time, and assignming resources to each operation (Kempf, Le Pape, Smith, & Fox, 1991; Waterman, 1986).	Not a specific research area.

Table 1.2 summarizes the concepts from an artificial intelligence and cognitive psychology perspective. After an extensive literature review we found no standardized measurement instrument of any construct in Table 1.2.

Measurement of Metacognition

Although there are no commercially available instruments for the measurement of metacognition, researchers in this field designed some approaches for their own research purposes. The techniques for measurement of metacognition in such empirical studies may be categorized into seven kinds. The following are examples of these methods.

Error Detection Paradigm.   This method is mostly used in the measure of reading comprehension monitoring. A short passage contains a single contradiction. The contradictory information is usually not in contiguous sentences. For example, one passage describes cave-dwelling bats that are deaf, and toward the end of the passage it is stated that the bats use echoes to locate objects (e.g., Walczyk & Hall, 1989). The subject is asked to detect the contradictions. If the inconsistent information goes undetected, it is assumed that readers have failed to monitor their level of comprehension adequately. The number of correct detections is used as an index of reading comprehension monitoring.

Self-Rating Scale.   The reading awareness interview was designed to assess children's awareness about reading in three areas: evaluating task difficulty and one's own abilities, planning to reach a goal, and monitoring process toward the goal. The interview contains scale items (Jacobs & Paris, 1987). For example, one monitoring item is "Why do you go back and read things over again?" with three scored choices: (a) *because it is good practice* (score = 1); (b) *because you didn't understand it* (score = 2); (c) *because you forget some words* (score = 0). Jacobs and Paris (1987) suggested that the scale is sensitive to developmental and instructional differences in children's metacognition about reading.

Questionnaire Inventory (Self-Report).   Learning strategy inventories use this form (e.g., Weinstein, Palmer, & Schultz, 1987). For example, the Learning and Study Strategies Inventory (Weinstein et al., 1987) includes the following categories of learning and study strategies or methods: (a) attitude and interest; (b) motivation, diligence, self-discipline, and willingness to work hard; (c) use of time management principles for academic tasks; (d) anxiety and worry about school performance; (e) concentration and attention to academic tasks; (f) information processing, acquiring knowledge, and reasoning; (g) selecting main ideas and recognizing important information; (h) use of support techniques and materials; (i) self-testing, reviewing, and preparing for classes; (j) test strategies and preparing for tests. The following is an example of an item for self-testing:

> After class, I review my notes to help me understand the information (1) *not at all typical of me,* (2) *not very typical of me,* (3) *somewhat typical of me,* (4) *fairly typical of me,* (5) *very typical of me.*

Think-Aloud Protocol Analysis.   Think-aloud protocol analysis asks subjects to vocalize their thinking processes while working on a problem. The data as a protocol are then coded according to a specified model for psychological analysis, which provides insights into elements, patterns, and sequencing of underlying thought processes. For example, subjects were asked to solve a LOGO programing problem. The statements from their think-aloud protocols during problem solving were categorized in the scheme of Sternberg's componen-

152, 112

tial intelligence model including these components: deciding the nature of the problem, selecting performance components, combining performance components, selecting a mental representation, allocating resources, monitoring solution, and being sensitive to external feedback (Clements, 1987).

Evaluating Relative Efficacy of Strategies Used.   In this technique, the presence or absence of metacognition is inferred by some sort of experimental manipulation. For example, the underlying assumption is that one aspect of metacognition is to monitor strategies for use in the regulation of ongoing cognitive activity. Two memory strategies were introduced to the subjects when they were asked to remember a list of vocabulary. Then, they were asked to evaluate the relative efficacy of the strategies according to their memory experience (Brigham & Pressley, 1988).

Behavior Observation.   It is expected that changes in problem-solving conditions would result in changes in cognitive processes that would be, in turn, reflected in overt behavior measures such as speed. The "data" in this case are not the subjects' cognitions, but their overt external behaviors. For example, subjects were requested to solve puzzles under reversible and irreversible conditions (in the irreversible condition, once a piece of the puzzle was placed on the working cardboard, it became fixed and could not be removed). Changing the problem-solving condition would cause children to increase the intensity and to decrease the speed of their solution approach. The change of actions was viewed as the evidence of monitoring and regulating the course of their problem-solving processes (Kluwe, 1987).

Awareness of Prerequisite Knowledge.   Tobias, Hartman, Everson, and Gourgey (1991) stated that metacognition depends on students' ability to examine the state of their declarative knowledge about different subjects and their capacity to distinguish between what they know and do not know. In an experiment, students were asked whether they knew or did not know a list of vocabulary words and were then given a vocabulary test containing the same words to determine the accuracy of their judgments. Validity was judged with a reading comprehension test.

Cognitive and Expert System
Taxonomies or Categories

To use a cognitive skill instrument in human benchmarking of an expert system, one primary task is to examine the possible overlappings between categories of expert systems and types of cognitive skill instruments. Only when a particular category or function of expert systems is identified to be parallel or not parallel to a particular type of cognitive skill instrument, can the decision be made that the particular instrument is legitimate to be used for norming the performance of expert systems on that of human beings, and that this human benchmarking approach is valid. Both general and specific approaches were used to accomplish this task. The general approach used two category systems of expert systems (Chandrasekaran, 1986; Shalin, Wisniewski, & Levi, 1988), whereas the specific approach employed an expert system (GATES). Their possible correspondences were examined based on the descriptions or definitions of the categories or functions of expert systems and cognitive skill instruments.

Our general approach used two category systems (or classifications/taxonomies) of expert systems (Chandrasekaran, 1986; Shalin et al., 1988). Their possible correspondences were examined based on the descriptions or definitions of the categories or functions of expert systems and cognitive skills. We examined the definitions of expert system categories and those of cognitive skills or factors an instrument is intended to measure to determine in what ways they are or are not parallel.

As shown in Table 1.3, Shalin et al.'s (1988) system categorizes expert

TABLE 1.3
Expert System Function Classification

Shalin, Wisniewski, & Levi (1988)

Classification
Matches the input features of an examplar of a class to a concept.

Interpretation
Constructs a coherent representation from classified objects.

Design
Arranges objects according to constraint on these objects.

Problem solving and planning
Arranges actions according to constraints on action sequences.

Chandrasekaren (1986)

Hierarchical classification
Organizes concepts in terms of their relations with the top-most concept having control over the subconcepts.

Hypothesis matching or assessing
Generates a concept, matches it against relevant data, and determines a degree of fit.

Abductive assembly
Assembles generated hypotheses into a composite hypothesis that best explains the data.

Hierarchical design by plan selection and refinement
Chooses a plan based on some specification; instantiates and executes parts of the plan, which in turn suggests further details of the design.

State abstraction
Predicts a state change when a proposed action may be executed.

---

systems according to their functions and knowledge requirements. These four categories are classification, interpretation, design, and problem solving and planning. They are described as hierarchically inclusive because the functions and the knowledge requirements for more complex expert system functions subsume requirements for less complex expert systems. Chandrasekaran's system (1986) identifies five critical functions or features called "generic tasks." The five critical functions, as shown in Table 1.3, are hierarchical classification, hypothesis matching or assessment, abductive assembly, hierarchical design by plan selection, and state abstraction. The "generic task" analysis seems to be more able to capture similar functions from different expert systems because it focuses on general functions of expert systems rather than specific tasks they perform, which is consistent with our approach in benchmarking expert systems. However, the function of human cognitive monitoring and expert system monitoring may differ in that human cognitive monitoring regulates on-line cognitive processes of problem solving, whereas a monitoring system does an on-line task such as weather monitoring, and so forth. It may be in this sense that Shalin et al. (1988) viewed the function of a monitoring system as similar to that of a classification system that functions so as to match the input features of an example of a class to a concept or its internal representation of the class. Thus, the functions of cognitive monitoring and an expert system monitoring something external are not the same thing.

In order to be clear about the meaning of monitoring, one needs to define the construct of self-monitoring in psychological terms. Our definition of this construct is a synthesis of the views of Weinstein and Mayer (1986) and Beyer (1988). We view self-monitoring as conscious and periodic self-checking of whether one's goal is achieved and, when necessary, selecting and applying different strategies. Thus, to self-monitor one must have a goal (either assigned or self-directed), and one must have a cognitive strategy to monitor (e.g., finding the main idea). Further, one needs a mechanism to know which strategy among competing strategies to initially select to solve a task and, further, when to change such a strategy if it is ineffective in achieving the goal. Finally, these processes must be conscious to the individual. These constructs of monitoring, planning, cognitive strategy, and awareness constitute our working definition of metacognition.

When self-monitoring, a problem arises when one's initial cognitive strategy is ineffective but there is no other strategy to select. The latter case is common with lower aptitude students. For example, many low aptitude students memorize information by repetition alone and have no other strategies to use when repetition is inefficient or ineffective. Thus, although conscious of failure, they have no other strategy to select and use (e.g., use of imagery). Surprisingly, in an expert system, the self-monitoring to assess efficacy of strategies used finds its best realization in the application of scheduling, rather than in monitoring systems.

The typical goal of an expert system scheduler is to assign an item to a specific time, location, and so on, without violating any constraints. The scheduler makes assignments according to some prescribed strategies such as the number of constraints assigned to the item or the distance among items. If an expert system's initial strategy fails to assign all the items, then the program must either relax its constraints or adopt a new and different strategy (Dhar & Ranganathan, 1990). Similar to the human example, a more sophisticated and complex expert system should be able to select a different scheduling strategy if its initial strategy proves ineffective.

The parallel analysis of cognitive and expert system categories suggests that the parallel relations between them are not simple. For the instance of classification, the induction and the logical reasoning tests from the Kit of Factor-Referenced Cognitive Tests (Ekstrom, French, & Harman, 1976) were chosen. The induction test measures the ability to form and try out hypotheses that will fit a set of data. It asks subjects to find concepts of classes that will group all the given objects into these classes. The nature of the task seems to be in correspondence with Shalin et al.'s definition of classification as matching the input features of an exemplar of a class to a concept. The logical reasoning test identifies the ability to reason from premise to conclusion. One task of the test requires the construction of hierarchical relations of classes—which is very matched to Chandrasekaran's hierarchical classification category—which is defined as organizing concepts in terms of their relations, with the top-most concept having control over the subconcepts. However, there is a subtle difference in that a classification activity for humans includes concept formation and concept identification (e.g., Vygotsky, 1978). Concept formation employs a bottom-up (forward chaining in artificial intelligence terminology) approach, whereas concept identification uses a top-down (backward chaining) approach. The classification function in an expert system application usually contains only the top-down approach (e.g., Brazile & Swigger, 1988).

In addition, the data from developmental psychology (e.g., Bruner, Olver, & Greenfield, 1966) show that classification of objects by function is viewed at a higher level than classification of objects by shape. But the same judgment may not be made in expert system applications. It may be inappropriate to say that a system doing shape classification is "smarter" than one doing function classification because both carry out the same function in a technical sense.

The analysis of the previous example suggests both the possibility to use cognitive skill instruments to norm expert system performance and the subtle distinctions between them.

Analysis of the GATES System

The expert system GATES was used for an analysis of the correspondence between the system's functions and Shalin et al.'s (1988) and Chandrasekaran's (1986) categories.

GATES is an expert system written in Prolog for gate assignment of airplanes at TWA's JFK and St. Louis air terminals (Brazile & Swigger, 1988, 1989). It is a production program with documentation and source code available. It was chosen as a target system for our human benchmarking approach for two reasons. First, the system has some features (e.g., monitoring functions) we were interested in for the benchmarking evaluation of expert systems. Second, and more important, we had good cooperative relationships with the developers of the system. This relationship permitted us to capture all the relevant documentation as well as the informal details about the system. Based on the aforementioned literature and expert judgment, we benchmarked the performance of the GATES system to a human standard using an empirical methodology.

GATES is a constraint satisfaction expert system developed to create TWA's monthly and daily gate assignments at JFK and St. Louis airports. Obtained from an experienced ground controller, the domain knowledge is represented in Prolog predicates as well as several rulelike data structures, including permission rules (the GATEOK predicate) and denial rules (the CONFLICT predicate). These two kinds of rules determine when a set of gates can or cannot be assigned to a particular flight.

The system uses the following procedures to produce monthly gate assignments:

1. Consider an unassigned flight that has the most constraints first (a set of FLIGHT rules).
2. Select a particular gate for a particular flight by using a set of GATEOK rules that have been arranged in some priority.
3. Verify whether the gate assignment is correct by checking it against a set of CONFLICT rules.
4. Make adjustments by relaxing constraints to have all flights assigned to gates.
5. After all assignments are made, adjust assignments to maximize gate utilization, minimize personnel workloads, and maximize equipment workload.

In summary, the GATES system monitors itself in three passes:

Pass 1: Use all of the constraints; try to schedule all the planes. With all the constraints in use, usually only about 75% of the planes get scheduled, and the system moves to Pass 2. (If successful in scheduling all planes, the system still goes to Pass 2.)

Pass 2: Relax the constraints so that all of the planes are scheduled, and move to Pass 3.

Pass 3: Put back the constraints in an attempt to optimize the schedule.

All passes recurse through Passes 1, 2, and 3. Thus, Passes 1, 2, and 3 act as a kind of subprogram or subprocedure that is called during every phase of the program. Pass 1 recurses only through Procedures 1, 2, and 3. Pass 2 performs Procedure 4, while it recurses through 1, 2, and 3. Pass 3 performs Procedure 5, while it recurses through Procedures 1, 2, and 3. Because Prolog uses recursion, it is difficult to separate these passes into separate procedures.

EXPERT SYSTEM APPLICATION
OF HUMAN BENCHMARKING

Our expert system human benchmarking methodology consists of 11 steps from the initial selection of an expert system to the final report documenting the process (see Table 1.4). As may be seen in Table 1.4, following selection of an expert system, one would classify the expert system as to its application. The possible applications are diagnosis, monitoring, planning, and scheduling. Then, one would classify the expert system within a computer science taxonomy.

Next, one creates an analogy, that is, the functioning of this expert system within a certain computer science taxonomic classification is matched onto a specific cognitive functioning of humans (e.g., monitoring). The analogous functioning is classified within a cognitive science taxonomy. Then, one selects or develops both cognitive and affective measures of human functioning on the task. Both process and outcomes are measured. Next, one selects an empirical design (qualitative or quantitative) and runs the empirical studies. The data are statistically analyzed, and then either existing norms are used or new ones are created. Finally, a report on the "intelligence" of the expert system is written.

The application of the general method to the specific case of GATES in this report is shown in Table 1.5. As seen in Table 1.5, we have instantiated all steps of the general method with one exception in that the last step of the general method, "report intelligence of expert system," was broken down into two sub-

TABLE 1.4
Expert System Human Benchmarking Methodology

Select expert system
Classify expert system as to its application
Classify within a computer science taxonomy
Create analogy
Classify within a cognitive science taxonomy
Select/develop measures of analogous functioning
Select empirical design
Conduct empirical studies with people
Analyze statistically
Use/create/norms
Report "intelligence" of expert system

TABLE 1.5
Expert System Human Benchmarking Methodology for "GATES"

General Methodology	Specific Example
Select expert system	GATES
Classify as to application	Scheduling
Classify within computer science taxonomy	Hierarchical design
Create analogy	Metacognition
Classify within a cognitive science taxonomy	Metacognition
Select/develop measures of analogous functioning	Thinking questionnaire
Select empirical design	Pilot study, descriptive study
Conduct empirical studies with people	This chapter
Analyze statistically	This chapter
Use/create/norms	To be done
Report "intelligence" of specific expert system	This chapter
Report "intelligece" of expert system application	To be done

steps: (a) report intelligence of the specific expert system (i.e., GATES) and (b) report intelligence of expert system application (i.e., scheduling). The second substep would require extensive and expensive empirical research. This chapter uses several studies as "proof of concept" of our human benchmarking methodology but does not provide enough information (i.e., norms) to report on the intelligence of the expert system classification.

GATES is classified as a scheduling system. Its computer science taxonomy is a hierarchical design by plan selection and refinement (Chandrasekaran, 1986). With respect to its analogy, we consider monitoring in GATES to be like metacognition in people (see Fig. 1.1). For the scheduling task, the system and people share the same goal: to assign all landed flights to available gates. Both the system and people follow the same constraints and rules to do the task. Following these restrictions, the system monitors itself in the three phases of scheduling. People use the same constraints to plan, monitor, and assess ongoing processes of scheduling. However, people are aware of their ongoing processes whereas the system is not.

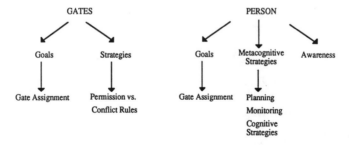

FIG. 1.1. An analogy for human benchmarking of GATES.

	Process	Outcome
People	a) Think-aloud protocols b) Metacognitive tests c) Affective measures	a) Number of correct gate assignments
System	a) Trace b) Categorized rules	a) Number of correct gate assignments

FIG. 1.2.   Process and outcome measures.

We modified an existing cognitive science taxonomy to classify monitoring processes (Beyer, 1988). We then developed measures for both process (e.g., for people, a questionnaire on metacognition) and outcome measures (see Fig. 1.2). Next, an empirical design was selected in which two scheduling problems of different difficulty levels were administered to junior college students, undergraduate and graduate students at a major university, and three experts who are airport ground controllers. The empirical study was run and the data were analyzed. The data of the pilot study and the junior college administration are reported in subsequent sections. The data on the two types of task for undergraduate and graduate students and the experts' performance are reported in O'Neil, Baker, Jacoby, Ni, and Wittrock (1990). A report on "intelligence" of a specific expert system, GATES, is contained in this chapter. The use or creation of norms then follows the empirical work. The final step remains to be done, as it requires extensive and, thus, expensive empirical work.

The function of norms is to help establish the status of the expert system's performance. Norms are determined in different ways in the psychological research literature. For example, on the Stanford–Binet intelligence test, norms are percentages passing a task by grade level. With standardized tests, the process varies. For instance, raw scores on the Iowa Tests of Basic Skills are converted to either developmental scores (grade equivalent, age equivalent, and standard scores) or status scores (e.g., percentile ranks) (Hieronymus, Hoover, & Lindquist, 1986). A good description of the norming process is provided by Petersen, Kolen, and Hoover (1989).

Scheduling Task Characteristics

The expert system GATES chosen for this research schedules airline flights to airport gates after the planes have landed. The goal of the task given to our human subjects was identical. There were certain constraints (rules and restrictions) in doing the task because there were different kinds of flights (domestic and international), various gates (that can have different planes scheduled to them), plane types (of different sizes and shapes), ferry flights (unscheduled flights in which an aircraft is transported with no passengers), and taxiways (roadways to the gates). Other constraints included turn times (the estimated amount of time it takes to unload/load passengers and clean the plane), separation times (time needed between two planes when they use the same gate), continuation flights (flights that arrive from one location and depart for a different location on the same day), and moves (moving of planes from gate to gate). The constraints were taken from the expert system GATES (Brazile & Swigger, 1988). The expert system GATES follows a plan in order to do the task. The plan includes a procedure or strategy of how to solve the task. General guidelines are given, such as which flights to assign first. There are three phases to the plan: The first phase is to try to complete the task with all the constraints. The second phase gives guidelines to relax the constraints in order to schedule all unscheduled flights. The third phase gives guidelines to include selected constraints again to the extent possible so that all the flights are still assigned.

In our studies, this plan, with slight modifications, was given to the subjects solving the task. The plan was modified in the following manner: (a) We added an explanation of how to use the plan that the expert system uses by default, such as *Pick out the constraints relevant to your task;* and (b) to reduce memory demands, the subjects also received a job aid, that is, a worksheet that was a table showing the flight times vertically and the gates with appropriate taxiways horizontally. The problem given to the subjects in the study included a list of incoming flights with the following information about each flight: flight destination (domestic or international), flight number, plane type, arrival time, and departure time. The task for the subjects was to assign the incoming flights to available gates without violating the constraints.

We also characterized various types of scheduling tasks by difficulty level as easy, medium, hard, and very hard tasks. The easy task reported in this chapter includes 15 flights, 10 gates, and 3 taxiways. Only domestic flights are given, and no continuation flights, moves, mobile lounges (predesigned flights that use a buslike vehicle to load/unload passengers), tow on gates (gates that need some type of equipment to place the plane at the gate), or ferry flights are given. The easy task can be solved using only the first phase of the plan, that is, by doing the task with all the constraints. A medium task includes 15 flights, 9 gates, and 5 taxiways. There are domestic and international flights, 4 continuation flights, and 4 moves. The second pass of the plan is needed in order to assign all the flights by relaxing some of the restrictions. A hard task includes 130–140 flights, 40 gates, and 5 taxiways. This task is a typical one for the expert system. All the constraints are needed to do the task, and all three phases of the plan are required. This task is the hardest task the computer can do successfully. A very hard task is similar to the hard task but one more flight is added that the computer is not able to schedule because too many constraints need to be relaxed. A very experienced flight assignment controller can do the scheduling of this very hard task.

Pilot Study

The goal of the pilot study was to test the human scheduling task materials for content, clarity, and administration time before administering the task to a larger sample of students. Further, we also pilot tested our psychological process instruments (a metacognitive skills questionnaire and a state worry scale). Finally, various scoring techniques were developed for errors on the scheduling task.

Subjects. Seven people served as subjects for the pilot study. The seven subjects included one professor, one secretary with a master's degree, two graduate students, and three undergraduate students. Both male and female subjects participated with a range of ages and status. All subjects had taken at least one domestic or one international flight within the last 5 years. Four of the subjects were paid for participating in the research, and two of the other three subjects were people working on the project but not familiar with the task. The principal investigator served as the remaining subject. The subjects can be characterized as highly motivated to succeed in the task.

Materials. The pilot study was conducted using both the easy and medium difficulty tasks. Most subjects performed either one task or the other, but some did both. All subjects received the worksheet, and all but the secretary received the plan. Table 1.6 indicates which subjects received what task and whether they received one or two tasks.

TABLE 1.6
Summary of the Pilot Study

Subjects	Easy Task		Medium Task	
	Errors	Time	Errors	Time
Secretary[a]	5	60 min	–	–
Professor	3	50 min	3	45 min
Graduate student 1	0	120 min	–	–
Graduate student 2	0	90 min	1	70 min
Undergraduate student 1	0	50 min	–	–
Undergraduate student 2	5	55 min	–	–
Undergraduate student 3	–	–	0	130 min

Note. Dashes indicate that the subject did not perform this task.
[a]No plan was provided.

Procedure.   The data for the pilot study were collected on an individual or two-person group basis. The subjects were given as much time as needed, and the experimenter kept track of the time. After the task was completed, the experimenter asked the subjects questions about the difficulty and clarity of the directions and the task. The subjects were also asked for the process they followed to complete the task.

Results and Discussion.   The data were checked and analyzed for correctness and time to complete the task. Table 1.6 also presents the results; the numbers represent the number of wrong gate assignments.

The secretary (our first pilot subject) performed the task without a plan. She had many problems doing the task. Thus, we decided that because the computer follows a plan to schedule the flights, human subjects should received a similar plan. Therefore, the rest of the pilot subjects received the plan with the task. The results of using the plan reduced the number of errors and time to complete the task(s). As may be seen in Table 1.6, the undergraduate student who did the medium difficulty task needed much more time to do the task because, we assume, of not having the experience of solving the easy task first. Both graduate students that performed the easy task solved it with no errors. Although one undergraduate student solved the task with no errors, the other undergraduate student had 5 wrong assignments out of 15 listed flights.

Junior College Study

The goal of this part of our research program was to administer the scheduling task to a larger group of students whom we assumed to be of average intellectual ability and to have low to medium motivation levels to succeed on the task.

Subjects.   Twenty-seven junior college students in an intact class participated in the study. Two of the subjects were non-English-speakers; therefore, their data were not included in the data analysis. The students were all members of an English literature class at a local junior college. Students in this literature class were predominantly persons who wished eventually to complete a bachelor's degree at a 4-year college. Both male ($n = 14$) and female ($n = 10$) students participated; one subject did not provide gender information. Students had different school majors and a range of GPA (2.3–4.0). All subjects had taken at least one domestic flight and most had taken international flights within the last 5 years. Two of the students had previous scheduling experience: one student in scheduling buses to the Hollywood Bowl, and another in political campaign projects. The students can be characterized as somewhat motivated but wanting to finish the task quickly.

Materials.   The materials used for the study include directions, a prequestionnaire, the scheduling task, and posttask questionnaires. The prequestionnaire was a biographical questionnaire asking the students for name, gender, age, education, scheduling experience, and flight experience. The scheduling task included a plan to follow to do the task, the constraints (restrictions) to use in doing the task, a worksheet and extra blank paper, and the scheduling task itself. The task and materials were the "easy task" that was pilot tested.

The posttask questionnaires included a five-item poststate worry scale (Morris, Davis, & Hutchings, 1981); a metacognitive thinking questionnaire developed by Harold F. O'Neil, Jr., in this program of studies; and some additional questions on the task. The worry questionnaire asked students to rate their attitudes or thoughts while doing the task using a 1-to-5-point scale. A sample item is "I felt I did not do as well on this task as I could have done." The subjects were asked to rate each item using the following scale: 1—*The statement does not describe my past condition;* 2—*The condition was barely noticeable;* 3—*The condition was moderate;* 4—*The condition was strong;* 5—*The condition was very strong.* The metacognitive thinking questionnaire's goal was to determine whether students were engaged in metacognition while doing the scheduling task. The students responded: 1—*Not at all;* 2—*Somewhat;* 3—*Moderately so;* or 4—*Very much so.* There were 26 items. The questions asked about students' planning, monitoring, cognitive strategy use, and awareness. A sample item for planning is "I explicitly planned my course of action."

Procedure.   The study was conducted in a regular, scheduled, English literature class of 55 min. The teacher had told the students they would be asked to participate in a study. The experimenters told the students the task involved scheduling airline flights to gates after they landed and that the task was part of a research study of comparing human performance to computer performance. The students were given 55 min to read the directions, answer the biographical questionnaire, and do the task using the constraints, plan, and worksheet. The experimenters marked the finish time of each student on the scheduling task and then gave each student who finished the task the posttask questionnaires.

Results and Discussion.   Because this study only used one group of students and only the easy task, our basic analysis is descriptive. The variables included in the analysis are the gate assignment performance, level of state worry, and metacognitive processes covered by the questionnaires. The errors made were subdivided into eight categories: separation time, wrong plane type, time conflict, Gate 4 not free when should be, taxiway conflict, unavailable gates, illegal moves, and missing assignment. These categories are shown in Table 1.7, and reflect violations of the constraints. Every wrong assignment was categorized to one of these categories. A single wrong assignment could potentially result in multiple categorization of violations but would only be considered one error in our scoring. A perfect score was 15 where all 15 flights were assigned to appropriate gates; every mistake resulted in one point taken off.

Table 1.8 summarizes the results including the variables: gate assignment

TABLE 1.7
Scheduling Task Error Categories

Variable	Definition
Separation time violation	Less than 20 min separation time at the same gate between one flight to another.
Wrong plane-gate matching	A flight with specific plane type assigned to a gate that is not proper for the plane type.
Time conflict	Two flights are assigned to the same gate at the exact same time.
Gate 4 violation	A flight using 747 plane type is assigned to Gate 3, and Gate 4 is not left open.
Taxiway violation	Less than 5 min separtion time between flights using the same taxiway.
Unavailable gate	A flight is assigned to a different gate.
Illegal move	A flight is moved from one gate to another although it is not supposed to.
Missing assignment	A landed flight is not assigned to any gate.
Total wrong assignment	Total number of errors made.

performance, GPA, scheduling experience, travel experiences. The mean of total correct assignments was 11.96 (80%), which means the average of wrong assignments was three errors. Most of the errors made were in the time conflict and taxiway conflict categories. Further, both questionnaires exhibited acceptable reliability in that alpha reliability for the state worry scale was .74 and alpha

TABLE 1.8
Mean, Standard Deviation, and Range of Variables in the Gate Assignment Task

Variable	N	Mean	SD	Range
Total correct assignments	25	11.96	2.09	8-15
GPA	22	3.18	.36	2.3-4.0
Schedule experience[a]	24	.12	.338	0-1
Time (min)	22	43.91	9.75	25-60
Domestic travel experience	25	3.24	1.42	1-5
International travel experience	25	2.08	1.12	1-5
Separation time violation	25	.20	.50	0-2
Wrong plane type	25	.36	.57	0-2
Time conflict	25	1.00	1.41	0-4
Gate 4 violation	25	.36	.70	0-2
Taxiway violation	25	.76	.78	0-2
Unavailable gate	25	.28	1.21	0-6
Illegal move	25	.04	.20	0-1
Missing assignment	25	.12	.60	0-3
Total wrong assignments	25	3.04	2.09	0-7
State worry[b]	23	9.61	4.23	5-20
State self-monitoring[c]	21	73.04	14.14	35-102

[a]0 = No, 1 = Yes.
[b]Alpha reliability = .74.
[c]Alpha reliability = .91.

reliability for the metacognitive skills questionnaire was .91. The alpha reliability of the state worry scale (Morris et al., 1981) has been reported as .81. Given our small sample size ($N = 23$), our alpha reliability of .74 is acceptable. Further, Morris et al. (1981) reported mean levels of state worry of 12.43 ($SD = 4.65$) in a testing situation. Thus, compared to the Morris et al. data, our students ($M = 9.16, SD = 4.25$) are not exhibiting high levels of state worry. These low levels of state worry were expected as our instructions focused on the task and not the individual. The state metacognition data indicated that students exhibited a moderate amount of metacognitive activity.

The data in Table 1.8 were also described correlationally in Table 1.9. The results showed that the total number of correct gate assignments is negatively correlated to state worry ($r = -.55, p = .006$), but positively to metacognition ($r = .46, p = .036$) and level of domestic travel experience ($r = .42, p = .035$). The correlation of metacognition with state worry is marginally significant ($r = -.39, p = .086$). In general, GPA, time, and scheduling experiences were not related to other variables. In contrast, domestic travel experience was positively related to total correct assignments ($r = .42, p = .035$), negatively related to state worry ($r = -.42, p = .046$), and positively related to state self-monitoring ($r = .51, p = .017$). Thus, students with more domestic travel experience performed better on the task, had less state worry, and showed increased metacognition. In general, successful gate assignment performance on the easy task was positively correlated to metacognition, but negatively correlated to worry. These data indicate that the higher the state worry score, the lower the performance, whereas the more active the thinking process, the better the performance.

There were three junior college students who completed the task without errors. Because one goal of the study was to compare human performance to expert system performance, the task was given to the expert system GATES to get the computer's assignment to compare to the human subjects' assignments.

A Comparison With the Expert System GATES

The easy and the medium difficulty tasks were given to GATES to solve so comparisons could be made to human solutions. The computer solved both tasks with no errors. The expert system correctly assigned all planes to appropriate gates within 7.5 s for the easy task. For the medium difficulty task, GATES correctly assigned all planes within 23 s. A trace of the steps taken by the computer was made and analyzed by Kathleen Swigger, the developer of the program, and by Anat Jacoby (a member of the benchmarking team). The computer solved each problem in three passes, but the number of rules that were activated in each pass varied between the easy and the medium difficulty tasks. The first pass tries to schedule all the flights to the available gates without violating the constraints. The schedule order used was big planes first, and then

TABLE 1.9

Correlations Among the Variables for Gate Assignment Performance (Junior College Students)

	1	2	3	4	5	6	7	8	9	10	11	12	13	14	15
1 GPA	1.000														
2 Time	-.0308 (.898)	1.000													
3 Schedule experience	.0580 (.798)	-.1429 (.526)	1.000												
4 Domestic travel exper.	.0700 (.757)	.0442 (.845)	.1875 (.380)	1.000											
5 Int'l travel experience	.4134 (.056)	.0316 (.889)	.1586 (.459)	.3289 (.108)	1.000										
6 Separation time violation	-.0052 (.982)	-.4857 (.022)	-.0948 (.659)	.0469 (.824)	-.1046 (.619)	1.000									
7 Wrong plane	.2204 (.324)	-.2124 (.343)	-.4191 (.042)	.0433 (.837)	.0184 (.930)	.3224 (.116)	1.000								
8 Time conflict	.1154 (.609)	.0915 (.685)	.2607 (.219)	-.2071 (.320)	-.0264 (.900)	.2946 (.153)	-.0518 (.806)	1.000							
9 Gate 4 violation	.0202 (.929)	.2200 (.325)	.0000 (1.000)	-.2159 (.300)	-.0384 (.855)	.0238 (.910)	-.2889 (.161)	.2946 (.153)	1.000						
10 Taxiway violation	-.3062 (.166)	.0101 (.965)	.3647 (.080)	-.1715 (.412)	-.0249 (.906)	-.0856 (.684)	-.2672 (.197)	.0757 (.719)	.0886 (.673)	1.000					
11 Illegal move	—	.0708 (.754)	.0788 (.714)	.2578 (.213)	.5456 (.005)	-.0833 (.692)	-.1319 (.530)	-.1473 (.482)	-.1071 (.610)	-.2033 (.330)	1.000				
12 Missing assignment	.0100 (.965)	.0250 (.912)	.0788 (.714)	-.1816 (.385)	-.0149 (.943)	-.0833 (.692)	-.1319 (.530)	-.1473 (.482)	-.1071 (.610)	-.2033 (.330)	-.0517 (.843)	1.000			
13 Total correct assignment	-.0832 (.713)	-.1414 (.530)	.2364 (.266)	.4236 (.035)	.0372 (.860)	-.3905 (.054)	-.1626 (.437)	-.7608 (.000)	-.4452 (.026)	-.1340 (.523)	.2032 (.330)	.0040 (.985)	1.000		
14 State worry	.0766 (.748)	.0182 (.939)	.0356 (.875)	-.4208 (.046)	.1430 (.515)	-.0009 (.997)	.2677 (.217)	.4001 (.058)	.2458 (.258)	.3667 (.085)	-.2376 (.275)	.0202 (.927)	-.5552 (.006)	1.000	
15 State self-monitoring	.3622 (.140)	.0606 (.811)	-.0774 (.746)	.5146 (.017)	.2964 (.192)	.0247 (.915)	-.1246 (.591)	-.4859 (.026)	-.3779 (.091)	-.0890 (.701)	.1451 (.530)	-.2277 (.321)	.4603 (.036)	-.3938 (.086)	1.000

international flights. The order by plane type is 747S, 747, 1011, 767, MD80, 727S, 727. Flights that have continuation flights are scheduled first. The easy task did not have international flights or continuation flights, so the system picked a domestic flight and tried to schedule it. Usually a few gates were tried before an appropriate gate was found. The computer always tried the gates in a specific order, so when a flight was picked, all the gates were tried until an appropriate one was found. All the flights were assigned to gates after the first pass in the easy task. The medium difficulty task had international and domestic flights and continuation flights, and not all the flights were scheduled at the end of Pass 1.

The second pass looks at all the unassigned flights from the first pass and gradually lowers the tolerances until all the flights are assigned. The order in which tolerances are lowered is (a) separation time between departure of plane and arrival of new plane, (b) taxiway tolerances, (c) placing planes that are turning at a certain gate. In the easy task, GATES did not have problems assigning all the flights, so the second pass executed few rules. In the easy task, the program entered the second pass, checked that all flights were assigned, and completed Pass 2. For the medium difficulty task, on the other hand, more rules were executed by GATES in the second pass because not all the flights were scheduled in the first pass, and tolerances needed to be lowered in order to schedule all the flights.

The third pass tries to rearrange and improve the schedule. There is a rule in the third pass that checks to make sure there are a minimum number of moves. The program goes through a recursive routine in a process to try and maximize the intervals. It starts with the smallest interval. It then looks to see if there is another plane at the same time that can be switched that has a larger interval. If it looks good, it saves that idea and looks to see if the second plane can be moved elsewhere, and so on. If everything works out, then the system ripples back and makes the moves. It does this until it cannot find a better schedule. Both easy and medium difficulty tasks had rules activated in Pass 3. Improvements were tried for the easy task but none were found. For the medium difficulty task, only during Pass 3 were all flights scheduled, and then improvements were made. Altogether, the medium difficulty task was harder, and more effort by the computer was required in terms of rules being activated, which resulted in more time to solve the problem (7.5 s for the easy task verses 23 s for the medium difficulty task).

SUMMARY

This chapter documented our strategy for human benchmarking of expert systems. We modified the human benchmarking methodology used for a natural language understanding system. The metacognitive skills literature was reviewed; however, specific standardized, metacognitive tests were not found, so one was developed. By relating metacognitive taxonomies and expert system taxonomies, we developed a general approach to facilitate comparison. The expert system GATES matched our needs and was chosen as an example for our development of human benchmarking methodology for expert systems. The general methodology was developed and a specific example with GATES was instantiated.

In general, the feasibility of using human benchmarking methodology to evaluate a specific expert system has been established. The "intelligence" of this application (GATES) is extremely high in narrow domain of scheduling planes to gates, as indicated by the superior performance of the expert system compared to that of junior college students. Norms would be needed to quantify "extremely high." O'Neil, Baker et al., (1990) discussed the specific design and results of an experiment using both easy and medium difficulty tasks in human benchmarking using GATES. Their results confirmed those reported in this chapter.

ACKNOWLEDGMENTS

The authors wish to thank other members of the UCLA human benchmarking group: Frances Butler and Merlin Wittrock.

This research was supported by contract number N00014-86-K-0395 from the Defense Advanced Research Projects Agency (DARPA), administered by the Office of Naval Research (ONR), to the UCLA Center for the Study of Evaluation/Center for Technology Assessment. However, the opinions expressed do not necessarily reflect the positions of DARPA or ONR, and no official endorsement by either organization should be inferred. This research also was supported in part under the Educational Research and Development Center Program cooperative agreement R117G10027 and CFDA catalog number 84.117G as administered by of Office of Educational Research and Improvement, U.S. Department of Education. The findings and opinions expressed in this report do not reflect the position or policies of the Office of Educational Research and Improvement or the U.S. Department of Education.

REFERENCES

Baker, E. L. (1974). Formative evaluation of instruction. In W. J. Popham (Ed.), *Evaluation in education* (pp. 531–585). Berkeley, CA: McCutchan.

Baker, E. L., & Lindheim, E. (1988, April). *A contrast between computer and human language understanding.* Paper presented at the annual meeting of the American Educational Research Association as part of the symposium, "Understanding Natural Language Understanding," New Orleans.

Baker, E. L., O'Neil, H. F., Jr., & Linn, R. L. (1990). Performance assessment framework. In S. J. Andriole (Ed.), *Advanced technology for command and control systems engineering* (pp. 192–213). Fairfax, VA: AFCEA International Press.

Baker, E. L., Turner, J. L., & Butler, F. A. (1990). *An initial inquiry into the use of human performance to evaluate artificial intelligence systems.* Los Angeles: University of California, Center for the Study of Evaluation/Center for Technology Assessment.

Benwell, N. (Ed.). (1975). *Benchmarking: Computer evaluation and measurement.* Washington, DC: Hemisphere Publishing.

Berry, D. C., & Hart, A. E. (1990). Evaluating expert systems. *Expert Systems, 7*(4), 199–207.

Beyer, B. K. (1988). *Developing a thinking skills program.* Boston, MA: Allyn & Bacon.

Brazile, R., & Swigger, K. (1988). GATES: An expert system for airlines. *IEEE Expert, 3,* 33–39.

Brazile, R., & Swigger, K. (1989). *Extending the GATES scheduler: Generalizing gate assignment heuristics.* Unpublished manuscript.

Brennan, R. (1983). *Elements of generalizability theory.* Iowa City, IA: ACT.

Brigham, M., & Pressley, M. (1988). Cognitive monitoring and strategy choice in younger and older adults. *Psychology and Aging, 3,* 249–257.

Bruner, J. S., Olver, R., & Greenfield P. (1966). *Studies in cognitive growth.* New York: Wiley.

Camp, R. C. (1989). *The search for industry best practices that lead to superior performance.* Milwaukee, WI: Quality Press (American Society for Quality Control).

Chandrasekaran, B. (1986). Generic tasks in knowledge-based reasoning. High-level building blocks for expert system design. *IEEE Expert, 1,* 23–30.

Clements, D. H. (1987). Measurement of metacomponential processing in young children. *Psychology in the School, 24,* 23–30.

Cochran, C., & Hutchins, B. (1991). Testing, verifying, and releasing an expert system: The case of history of Mentor. In U. G. Gupta (Ed.), *Validating and verifying knowledge-based systems* (pp. 304–308). Los Alamitos, CA: IEEE Computer Society Press.

Cronbach, L., Gleser, G., Nanda, H., & Rajaratnam, N. (1972). *The dependability of behavioral measurements: Theory of generalizability for scores and profiles.* New York: Wiley.

Dhar, V., & Ranganathan, N. (1990). Integer programming vs expert systems: An experimental comparison. *Communications of the ACM, 33*(3), 323–336.

Dillingham, J., Tam, G. W., & Kotras, T. V. (1988). *Prototype expert system container ship stowage planning: User's manual* (Rept. No. MA-RA-840-88026). San Diego, CA: JAYCOR.

Ekstrom, R. B., French, J. W., & Harman, H. H. (1976). *Manual for kit of factor-referenced cognitive tests.* Princeton, NJ: Educational Testing Service.

Flavell J. (1981). Cognitive monitoring. In W. Dickson (Ed.), *Children's oral communication* (pp. 35–60). New York: Academic Press.

Geissman, J. R., & Schultz, R. D. (1988). Verification and validation of expert systems. *IEEE Expert, 3,* 26–33.

Guralnik, D. B. (Ed.). (1984). *Webster's new world dictionary.* New York: Simon & Schuster.

Hayes-Roth, B., & Hayes-Roth, F. (1988). A cognitive model of planning. In A. Collins & E. E. Smith (Eds.), *Readings in cognitive science: A perspective from psychology and artificial intelligence,* (pp. 496–513). Mateo, CA: Morgan Kaufmann.

Hayes-Roth, B., & Hayes-Roth, F. A. (1979). A cognitive model of planning. *Cognitive Science, 3,* 275–310.

Hayes-Roth, F., Waterman, D. A., & Lenat, D. B. (Eds.). (1983). *Building expert systems.* Reading, MA: Addison-Wesley.

Hieronymus, N. A., Hoover, H. D., & Lindquist, E. F. (1986). *Iowa Tests of Basic Skills (ITBS): Teacher's guide.* Chicago, IL: Riverside Publishing.

Hollnagel, E. (1989). Evaluation of expert systems. In G. Guida & C. Tasso (Eds.), *Topics in expert system design: Methodologies and tools* (pp. 377–416). New York: Elsevier Science Publishers, North-Holland.

Hudson, D. L., & Cohen, M. E. (1984, December). *EMERGE, a rule-based clinical decision making aid.* Paper presented at the IEEE Computer Society, First Conference on Artificial Intelligence Applications, Denver, CO.

Hushon, J. M. (1988). The feasibility of constructing an expert system to assist first responders to chemical emergencies. (Doctoral dissertation, The George Washington University, 1988). *Dissertation Abstracts International, 49,* 1300A.

Jacobs, J., & Paris, S. G. (1987). Children's metacognition about reading: Issues in definition, measurement, and instruction. *Educational Psychologist, 22,* 255–278.

Kempf, K., Le Pape, C., Smith, S., & Fox, B. (1991). Issues in the design of AI-based schedulers. *AI Magazine, 11*(5), 37–45.

Kluwe, R. H. (1987). Executive decisions and regulation of problem solving. In F. E. Weinert & R. H. Kluwe (Eds.), *Metacognition, motivation and understanding* (pp. 31–64). Hillsdale, NJ: Lawrence Erlbaum Associates.

Kopcso, D., Pipino, L., & Rybolt, W. (1988). A comparison of the manipulation of certainty factors by individuals and expert system shell. *Journal of Management Information System, 5*(1), 66–81.

Letmanyi, H. (1984). *Assessment of techniques for evaluating computer systems for federal agency procurements* (NBS-SP-500-113). Washington, DC: National Bureau of Standards (DOD), Institute for Computer Sciences and Technology.

Liebowitz, J. (1985, December). *Evaluation of expert systems: An approach and case study.* Paper presented at the IEEE Computer Society, Second Conference on Artificial Intelligence Applications, Miami, FL.

Mayer, R. E., & Wittrock, M. C. (in press). Problem solving and transfer of knowledge. In D. C. Berliner & R. C. Calfee (Eds.), *Handbook of educational psychology.* New York: Macmillan.

Mental measurements yearbook. (1985–1989). Lincoln, NE: University of Nebraska Press.

Moray, N. (1986). Monitoring behavior and supervisory control. In K. R. Boff, & J. P. Thomas (Eds.), *Handbook of perception and human performance* (pp. 40-1–40.51). (Vol. 2). New York: Wiley.

Morris, L. W., Davis, M. A., & Hutchings, C. H. (1981). Cognitive and emotional components of anxiety: Literature review and a revised worry-emotionality scale. *Journal of Educational Psychology, 73*(4), 541–555.

O'Neil, H. F., Jr., & Baker, E. L. (1987). *Issues in intelligent computer-assisted instruction.* Los Angeles: Center for the Study of Evaluation, University of California.

O'Neil, H. F. Jr., Ni, Y., & Jacoby, A. (1990). *Literature review: Human benchmarking of expert systems* (Report to DARPA, Contract No. N00014-86-K-0395). Los Angeles: University of California, Center for the Study of Evaluation/Center for Technology Assessment.

O'Neil, H. F., Jr., Baker, E. L., Jacoby, A., Ni, Y., & Wittrock, M. (1990). *Human benchmarking studies of expert systems* (Report to DARPA, Contract No. N00014-86-K-0395). Los Angeles: University of California, Center for the Study of Evaluation/Center for Technology Assessment.

Parasuraman, R. (1986). Vigilance, monitoring and search. In K. R. Boff, & J. P. Thomas (Eds.), *Handbook of perception and human performance* (Vol. 2, pp. 43-1–43-49). New York: Wiley.

Petersen, N. S., Kolen, M. J., & Hoover, H. D. (1989). Scaling, norming and equating. In R. L. Linn (Ed.), *Educational measurement* (3rd ed., pp. 221–262). New York: Macmillan.

Scriven, M. (1967). The methodology of evaluation. In R. W. Tyler, R. M. Gagne, & M. Scriven (Eds.), *Perspectives of curriculum evaluation. AERA Monograph Series on Curriculum Evaluation, No. 1* (pp. 39–83). Chicago: Rand McNally.

Shalin, V. L., Wisniewski, E. J., & Levi, K. P. (1988). A formal analysis of machine learning systems for knowledge acquisition. *International Journal of Man-Machine Studies, 29,* 429–446.

Sharma, R. S., & Conrath, D. W. (1992). Evaluating expert systems: The socio-technical dimensions of quality. *Expert Systems, 9*(3), 125–137.

Shortliffe, E. H. (1976). *Computer-based medical consultations: MYCIN (Artificial Intelligence Series).* Elsevier, NY: Elsevier Computer Science Library.

Sternberg, R. J. (1985). Cognitive approaches to intelligence. In B. B. Wolman (Ed.), *Handbook of intelligence: Theories, measurements and applications* (pp. 59–118). New York: Wiley.

Sviokla, J. J. (1990). An examination of the impact of expert systems on the firm: The case of XCON. *Management Information System Quarterly, June,* 127–140.

Tobias, S., Hartman, H., Everson, H., & Gourgey, A. (1991, August). *The development of a group administered, objectively scored metacognitive evaluation procedure.* Paper presented at the annual convention of the American Psychological Association, San Francisco.

Tonn, B. E., & Goeltz, R. T. (1990). Psychological validity of uncertainty combining rules in expert systems. *Expert Systems, 7*(2), 94–101.

U.S. Department of Defense. (1987). *Defense system software development* (and companion documents, DOD-SDD-2167A). Washington, DC: Author.

Vygotsky, L. S. (1978). *Language and thought.* Cambridge MA: Harvard University Press.

Walczyk, J. J., & Hall, V. C. (1989). Is the failure to monitor comprehension an instance of cognitive impulsivity? *Journal of Educational Psychology, 81,* 294–296.

Waterman, D. A. (1986). *A guide to expert systems.* Reading, MA: Addison-Wesley.

Weinstein, C. E., & Mayer, R. F. (1986). The teaching of learning strategies. In M. C. Wittrock (Ed.), *Handbook of research on teaching* (3rd ed., pp. 315–327). New York: Macmillan.

Weinstein, C. E., Palmer, D. R., & Schultz, A. C. (1987). *LASSI. The Learning and Study Strategies Inventory.* Clearwater, FL: H & H Publishing.

Weiss, S. M., & Kulikowski, C. A. (1988). *Empirical analysis and refinement of expert system knowledge bases.* Rutgers, NJ: The State University, New Brunswick, Center for Expert Systems Research.

Wigdor, A. K., & Green, B. F., Jr. (Eds.). (1991). *Performance assessment for the workplace* (Vol. 1). Washington, DC: National Academy Press.

Yu, V. L., Fagan, L. M., Bennett, A. W., Clancey, W. J., Scott, A. C., Hannigan, J. F., Buchanan, B. G., & Cohen, S. N. (1984). An evaluation of MYCIN's advice. In B. G. Buchanan & E. H. Shortliffe (Eds.), *Rule-based expert systems: The MYCIN experiments of the Stanford heuristic programming project* (pp. 589–596). Reading MA: Addison-Wesley.

2 Machine Vision: Metrics for Evaluating

Josef Skrzypek
University of California, Los Angeles

The long-term goal of the machine vision effort is to synthesize a general purpose system that can perceive and understand images of an unconstrained environment in "real-time." Our near-term goal is to understand how to specify a general purpose vision (GPV) system modeled after human vision. We assume that all of the visually guided human behavior can be realized in terms of a finite kernel of visual tasks. We attempt here to (a) identify these tasks, (b) translate them into routines and algorithms, (c) relate these routines to functional modules as constrained by information from neuroscience, (d) specify connectivity between modules, (e) synthesize functions computed by the modules in terms of neural structures.

We begin with the attempt to discern the definition of the general purpose vision system from a selection of existing computer vision systems. This is followed with the rationalization of using human vision as a model for general purpose vision systems and how to translate visual task identified through psychological experiments into computational modules of neural networks. Finally, we analyze in more detail the components of a visual system necessary for specification of general purpose vision systems.

A NEW PROPOSAL FOR A GENERAL PURPOSE VISION MACHINE IMPLIES ABILITY TO EVALUATE EXISTING SYSTEMS

Many different vision systems, realized or proposed, aspire to become a general purpose vision system. However, the definition of "general purposeness" is lacking, and it is not clear which visual tasks are generic to general purpose

47

vision. It is not even clear what should be a complete list of goals for the general purpose vision system. In absence of such definition, it is very difficult to derive common metrics for evaluation of the various systems. And without evaluation, it is difficult to propose new, improved, and more general systems. To define generality we need to evaluate current systems, and to evaluate them we need to have a common definition of generality.

Every existing machine vision system has been designed for a specific purpose. This means they cannot perform the same perceptual or categorization tasks. Consequently, evaluating their performance based on a fixed set of input images is almost impossible. It does not make much sense to evaluate vision systems using such techniques as figure-of-merit (FOM) (Gotlieb, 1985); figure-of-merit derived from weighted combinations of measures such as speed, reliability and accuracy are not likely to help in deciding which visual tasks are important or generic. After all, using a figure-of-merit to compare a system designed to locate three types of industrial plants in a highly constrained environment with another system designed to locate a camouflaged target in an unconstrained environment is meaningless. Such measures might be useful in the image-processing domain where, for example, convolving an image with some kernel is a well-defined procedure, but in vision the result of using figure-of-merit for general evaluation would most likely be misleading. Therefore, one of our research goals is to enumerate and categorize all visual tasks that a general purpose vision system must be able to perform. This might lead to a framework for evaluating progress in machine vision systems, useful not so much to judge other systems, but rather to discover what each system has proposed and where to direct future efforts.

The remainder of this chapter begins with a cursory review of 15 machine vision systems developed in the past in order to elucidate the definition of "general purpose" as well as all possible categories along which machine vision systems may be evaluated. This analysis is also intended to search for common, fundamental computational principles used by the different systems that we could embed into our general vision system. In the following section we attempt to justify the use of the human visual system as a model for general purpose vision systems. Next, we discuss the problem of comparing the visual performance of humans and machines, and which visual tasks can be used to measure the obvious gap. The chapter concludes with an outline of a proposed general purpose machine vision system derived from integrated analysis of current data in neurosciences.

REVIEW OF SELECTED COMPUTER VISION SYSTEMS

In an attempt to elucidate principles comprising the definition of a general purpose system, we first analyzed 15 systems built during the last 15 years (Table 2.1). We have based our analysis on five dimensions: (a) image attributes,

TABLE 2.1
Comparison of Systems

Principal Investigators	Image Attributes[a]	Perceptual Primitives[b]	Knowledge Base[c]	Object[d]	Representation[e]	Control Strategy[f]	Implementations[g]
Ballart, Brown, & Feldman (1978)	1, 2	1, 2	1	1, 2	1	1	2b
Hanson & Riseman (1978)	3a, 2, 3	1, 2	1	1, 2	1d	1, 2	1
Nagae & Matsuyama (1978)	1	2	1	1	1a	1, 2	1
Rubin (1978)	2, 3	2	1	3	1a	1, 2	3
Shirai (1978)	1	1, 1a	1	4	4	1, 2	2
Ohta (1980)	3a, 2, 3	2		1	1a	1, 2	1a
Brooks (1981)	1	2a, 1b		5	2		4
Nevatia & Price (1982)	1, 3a	3	1, 2	1, 2	1b		3
Shneier, Lumia, & Kent (1982)	3a, 1, 3	1, 1c	1	5a	5	1a, 2a, 2b, 3	4
Bolles, Howard, & Hannah (1938)	1	2	1, 3	5a	5	1, 2	4
Levine & Nazif (1984)	3a, 3b, 3c, 2	1, 2, 2b		1, 2	3	1a, 2a	4
Herman & Kanade (1984)	1, 1a, 4	1e, 2c, 1d	1	6	1c	2	3
McKeown (1985)	2, 3, 4	1, 2		3a	7	1, 2	1
Tsotos (1985)	1, 5	1f	1	4a	2	1a, 2a, 2b, 3	5
Davis & Kushner (1987)	1, 1a, 3, 5	1, 2	1	3b	6	1a, 2b, 3, 2a	2a

[a] Edges = 1; Point = 1a; Texture = 2; Color = 3; Intensity = 3a; Hue = 3b, Saturation = 3c; Depth = 4; Motion = 5.

[b] Line segments = 1; Curves = 1a; Ribbons = 1b; Corners = 1c; Junctions = 1d; Verticies = 1c; Markers = 1f; Regions = 2; Ellipses = 2a; Areas = 2b; Faces = 2c.

[c] Domain constrained = 1; Context constrained = 2; Location constrained = 3.

[d] Regions = 1; Lines = 2; 3D map = 3; Airports = 3a; Roads = 3b; Feature values = 4; Heart marker = 4a; Generalized cones = 5; Parts = 5a; 3D wireframe =6.

[e] Constraints graph = 1; Relational graph = 1a; Schematic network = 1b; Structure graph = 1c; Schemas = 1d; Frames = 2; Rules =3; Procedures =4; 3D CAD model =5; ALV terrain model =6; MAPS data base = 7.

[f] Top-down = 1; Data-driven control = 1a; Bottom-up = 2; Goal-driven control =2a; Model-driven control = 2b; Temporal-driven control = 3.

[g] Rule-based system = 1; Production system = 1a; Monitor program = 2; Vision executive = 2a; User executive program = 2b; Procedural = 3; Prediction system = 4; Frames = 5.

(b) perceptual primitives, (c) knowledge base, (d) object representation, and (e) control strategy. For the most part, we tried to use only systems that have been either proposed and partially realized or completely built.

Most of the systems use one or two image attributes such as edges and perhaps color. Some of them use higher level attributes, such as texture, and a few of the systems use motion. Image attributes can be defined as the most basic elements of pictorial representation that carry nonredundant information. By independent information we mean, for example, that color information cannot be obtained from texture or motion. It is clear that a general purpose machine vision system requires the exploitation of all of the image attributes (Table 2.1).

Perceptual primitives represent the second comparison dimension. A working definition of a perceptual primitive might be the abstraction of image attributes into higher level data structures. This abstraction may employ the use of Gestalt laws of organization (Wertheimer, 1938), such as closure, similarity, proximity, collinearity, common fate, and soon. Gestalt effects can be considered as mechanisms that group regions based on the idea of uniformity. Analyzed systems use very simplistic perceptual primitives that display very little abstraction and are intimately related to the original image attributes. Two of the most often used primitives are lines and region, directly related to contours. Most often, image attributes are simply integrated into new data structures. None of the systems uses more abstract perceptual primitives such as illusory contours.

A third way of looking at these systems is to examine the type of knowledge they use and the way the knowledge is represented and manipulated. In all of the systems, knowledge is constrained to a very narrow application domain. It is not immediately obvious how to compile all of the knowledge that relates to unconstrained environments.

We also examined the use and representation of objects. All of the systems use objects almost directly related to the very lowest levels of image representation. It would seem that some more complex (symbolic) representations, abstracted from combinations of those primitives, would be more beneficial. All of the available representational schemes such as graphs, frames, production rules, 3D CAD models, and generalized cones are used in various systems. For example, Hanson and Riseman's (1978) VISIONS schema structure provides a rich symbolic mechanism for the hierarchical representation of objects. In their method, an object could range from being an urban schema, which is composed of lower level subschemas such as house and road schemas, to simply a car schema. This type of rich symbolic image data structure is surely required for the development of a general purpose vision system, although the implementation may vary significantly.

Finally, we looked at the control strategy that is used to process the data within those systems. In most cases, both bottom-up and top-down strategies are used, but only a few systems have incorporated temporal aspects of the environment or goal-driven strategies to process the information.

It is extremely difficult to derive detailed conclusions from our preliminary comparisons of existing systems. In the absence of a common definition and specification for all these systems, and the lack of a good definition for general purpose vision, comparisons between these specialized systems can be cursory at best. Clearly, these systems do not use all of the available information from the early stages of processing.

The knowledge domain in all of the systems is highly constrained. The high-level vision components in all examined systems are rather weak and very much ad hoc. All of the high-level processes are derived from established artificial intelligence concepts (see Ballard & Brown, 1982) that have been borrowed from natural language processing research. It is not clear that processes underlying the higher levels of vision are identical to processes involved in language under-standing; most probably this is not the case. Although all systems are (implicitly or explicitly) able to perform object recognition tasks, the evaluation of this performance by the inventors is inconsistent or completely lacking. Some sys-tems can operate only on presegmented images of one object only. Others can "interpret" multiple objects but only if certain conditions are satisfied. These might include well-controlled illumination of the scene, an object placed in a particular position, object models developed independently of visual input, and so on. In general, the task of object recognition is not defined and only very simple objects are considered. For the most part, temporal aspects of the sensory environment have not been addressed by these systems. Finally, although every system starts out with the goal to build a general purpose vision system, either explicitly or by extension—for example see Witkin and Tenenbaum (1983)—little attempt was made to define the meaning of a general purpose system, much less the development of such a system. Perhaps this difficulty lies in the fact that we do not know what a general vision system is and we have never attempted to define strictly what is vision.

None of the machine vision systems presented in the previous section has the capability to perceive and understand images in an unconstrained environment. It is conceivable that many problems could be resolved if we had a specification for a general purpose system. Attempts to build systems using physical laws, such as the laws governing image formation, have proven to be intractable. Another promising approach to ill-posed vision problems is to reformulate them into problems that either have already been solved or for which a solution can be discovered. The reformulation can be realized only through the introduction of numerous constraints, thereby making the problem more tractable. However, the constraints often change the problem beyond its original specification. Although this approach has met with some success in low-level visual tasks, such as edge detection (Torre & Poggio, 1984), its application to higher level visual tasks, such as shape from shading (Ikeuchi, 1981), has failed.

All of the realized systems are unique, special case systems that work well in their dedicated application domain. Hence, it is very difficult or impossible to have a meaningful comparison of a system against its competitors or different approaches (see also Besl & Jain, 1985). It is even difficult to decide if all of the domains that have been selected are the right ones, or even significant ones, or how they contribute to our overall desire to understand and implement a general purpose system. In this sense the review of existing machine vision systems did not help in defining general purpose vision. However, it is expected that machine vision systems under development for operation in unconstrained environments, such as the autonomous land vehicle research effort, will contribute significantly to understanding these problems.

HUMAN VISION AS A MODEL OF GENERAL PURPOSE VISION

What can be gained from the analysis of the human visual system? Clearly, it serves as an existence proof of a general purpose vision system, capable of adapting to the requirements posed by the unconstrained environment in which we live. Computer vision literature is full of examples where the limits for machine vision performance are modeled after the limits of human vision (Besl & Jain, 1985). By analyzing the human visual system, the only vision system that works, we can have a better understanding of what vision is and what are the critical visual tasks that must be performed by a general purpose system. Along this line, the analysis of visual deficits in the human visual system may shed some light on the functional organization and the neural mechanisms underlying the performance of particular visual tasks.

Interdisciplinary Approach to Computer Vision

Progress in computer vision can benefit from interdisciplinary research contributions from neurosciences, psychology, optics, computer science, and electrical engineering. Our primary interest is in the phenomenon of visual recognition, which can be defined as the ability to extract from the two-dimensional retinal input all meaningful patterns of light and to interpret them as representation of objects, scenes, and their properties.

One influential example of elucidating potentially useful principles for machine vision by constructing computational models of low-level visual functions of biological systems is due to Marr (1982). His first work in the area of vision was an attempt to develop theory based on neurophysiology of the retina (Marr, 1976). The approach was to use the physiology and anatomy of the neural structures in the retina to constrain the mathematical model. In his later work this philosophy evolved to viewing perception as an information-processing problem where algorithms must be invented before attempting to understand neuronal structure.

Our approach is first to enumerate all visual tasks as high-level processes that comprise the phenomenon of unconstrained perception and then to specify the resulting functional modules in terms of underlying neuronal structures constrained by knowledge from neurophysiology. We assume that the early stages of visual processing consist of functional modules that decompose input image into features. Modules have properties derived from physiological studies: discernible boundaries; their internal architecture is subject to laws of adaptation and developmental plasticity; they can have local short-term memory; and they interact with others, both laterally and vertically, and therefore can interact with some higher level resources such as attention. This specification of modules is contrary to the one suggested by Fodor (1983), where clearly unidirectional links of the hierarchy prevent access from the higher to lower levels.

There are many ways to synthesize models of the brain functions. One approach, based on mathematical methods, tries to understand from the mathematical viewpoint the capabilities of neural networks that might have no correspondence in biological systems. Another way of theorizing about the brain function is exemplified by artificial intelligence. This is a top-down approach, where hypothesized brain functions are simulated as "intelligent behavior" in the form of a computer program without reference to any underlying neural mechanism. Our approach is different from the previous two methods in that our neural net models of specific brain functions adhere to current physiological and psychophysical data. Although our model is not identical in details to reality, we can learn a lot from achieving functional equivalence (Harmon & Lewis, 1966). For example, computationally, there are two aspect of image segmentation: (a) physical and geometric constraints of the world and (b) the specification of the computing substrate. The constraints must be discovered and posed in some mathematical formalism in order to be solved. However, the selection of the formalisms is constrained itself by the underlying computing architecture. In this sense our approach is different from the one popularized by Marr (1982), who advocated an algorithmic approach to vision; global brain phenomena, captured by psychology need only to be model by mathematical formalisms, and once in this form they can be computed by any substrate.

Because our models are expressed in terms of specific neural architectures, we can characterize our approach as mostly bottom-up. The key principle here is that, given the behavior of components and how they are interconnected, it should be possible by simulation to study and to specify a global behavior of the system. Following general rules about local connectivity, it is possible to achieve the emergence of order out of seemingly chaotic organization (Linsker, 1986). One benefit of this bottom-up approach is that by explicitly specifying the modeling paradigm we are forced to address many issues that would otherwise be neglected. By studying computer simulations of the model, we can predict neural phenomena of interest to neurophysiologists, and we can investigate how faithfully the model generates emergent global functions that are analogous to psychophysical data. This will help to extract most useful principles for synthesizing computer vision.

It is conceivable that complexity of the neural structure in the human perceptual function does not have the same meaning as in machine perception. Namely, a biological vision system might use a heuristic solution in the form of structural additions to genetically inherited architectures that are not decomposable into simpler computing units. Furthermore, our strategy of "Function from Structure" (Skrzypek, Mesrobian, & Gungner, 1989) when specifying general purpose vision is subject to the basic problem of neurosciences, namely, how to relate the neuronal computing substrate to the function it performs. Considering that the brain consists of billions of neurons, each with perhaps thousands of synapses, arranged into varying, task-dependent architectures, it is doubtful that neuroscience will be able to study in a deterministic manner every synapse. Therefore, modeling studies by computer simulations seem to be all the more significant and perhaps the only reasonable approach to the problem.

Can we synthesize a general purpose machine vision system without taking advantage of knowledge transfer from psychology or neurophysiology of vision? The slow progress of the past 30 years suggests it would be an extremely difficult undertaking. Which aspect of biological knowledge would be most useful? Phenomenological description of visual processes at the global level as offered by psychology, or detailed outline of local neural architectures as described by neurophysiology, or both? One argument posed by psychology (Hochberg, 1987) is that machine vision systems should not be designed to emulate human vision, but it would help if machines *knew* how people see; they should not suffer from visual inconsistencies due to depth reversals, illusions, or apparent motion. The implication is that such illusions represent unwanted byproducts of a very complex and general purpose system. However, consider the role played by illusory (subjective) contours when we recognize an object presented as an incomplete figure; illusory contours aid us in completing the missing data and in fact neurons have been found that specialize in detecting subjective contours (Baumgartner, von der Heydt, & Peterhans, 1984). This implies that our ability to perceive illusory contours is not just a result of visual inconsistencies, but rather a purposeful feature of the system, designed to aid in our perception of the environment.

Another implication of Hochberg's viewpoint is that we should keep developing vision systems specifically tailored to handle the particular tasks. In general, it is virtually impossible to apply these highly specialized systems to new tasks without major redesign. In the past 30 years, this approach has resulted in many excellent dedicated imaging systems, but it did not provide much useful knowledge on how to build a more robust vision system; knowing details of an insect vision specialized for detecting ultraviolet radiation in navigation tasks might not help much in understanding problems of higher level vision in human systems.

Human Visual Tasks

The list of the manifest properties of a general purpose system has never been compiled in the past. Which characteristics of perception would qualify the system for successful operation in an unconstrained environment? Clearly, it must be able to handle a wide assortment of visual tasks, some of which we have considered in previous sections. There are no rigorous studies that attempt to enumerate and define all visual tasks. Consequently, this is one of our goals in this project.

We assume that the human perceiver represents a general purpose vision system for which we are able to enumerate most of the visual tasks. Some of these tasks might be very difficult for current implementation of machine vision systems, others might be trivial. We pose that in human perception, there exists a kernel of visual tasks, a subset of which underlies most of the visually guided behavior in an unconstrained environment. Having this list of visual functions would simplify our task to develop specifications of a general purpose machine vision.

One compilation of visual tasks has been developed by the Educational Testing Service for the measurement of childhood cognitive development. The underlying mechanisms that permit human vision to succeed in such tasks are not completely understood. This is an area of active investigation within neuropsychology and cognitive psychology. Knowing these mechanisms would perhaps allow us to define some of the underlying primitive visual functions that could be transferred to machine vision. The problem is that we need to select tasks that allow meaningful transfer from human performance to machine vision.

The Hidden Figures Test CF-1 (Ekstrom, French, Harman, & Dermen, 1976) is an example of a visual task that requires the participant to find a given shape in some complex picture (the shape undergoes no rotation or size changes when placed in the picture; Fig. 2.1). This appears difficult for us because of the cluttered background. Using perhaps very high-level processes of visual cognition, we can usually solve this visual task after staring at it for while. In the process of looking at it, we probably employ such subtasks as boundary tracking, symmetry finding, matching, and so on. A computer program, on the other hand, using very low-level processes such as template matching on a run-length encoded image could complete the task in a fraction of a second. Therefore, a

FIG. 2.1. Hidden figure task. Find predefined shape in the clutter image.

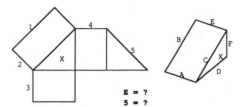

FIG. 2.2.   Surface development task.

comparison between human and machine performance based on this task entails computation at different levels of processing space and is not be very revealing.

Another example test is called the Surface Development Test VZ-3 (Ekstrom et al., 1976). In this case, the task is to fit a given surface together into a line drawing representation of the three-dimensional object (Fig. 2.2). Some tasks of this type are easy, whereas others are very difficult and might involve very different problems of selecting and representing models, manipulating internal models of complex three-dimensional objects, and finally matching the models against the data. Some of these problems, such as geometric reasoning, pose a great challenge to artificial intelligence (AI) in general, although they are relatively routine for humans.

Another task to consider is the Gestalt Completion Test CS-1 (Ekstrom et al., 1976), which might be closely related to segmentation. In this task, a subject is given incomplete data such as that presented in Fig. 2.3. The subject must determine what the data represents. Problems posed by this test include how to match the incomplete data against a set of models that one has acquired over time. There are many difficult questions to address. What is a model? How are models represented? How much data is needed? How are models matched against incomplete data? How do we implement something like illusory contours that would help in the performance of this type of task? Do we first complete the data by filling in the illusory contours? Biological vision has evolved specialized neurons that sense illusory contours after proper context has been analyzed. On the other hand, computer vision has yet to demonstrate an algorithm that even can detect the possibility of a contour completion in absence of data.

We have begun developing a prototype of a visual task sourcebook. The selected visual tasks range in difficulty and in their association with low- or high-

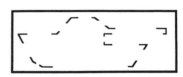

FIG. 2.3.   Contour completion task.

level visual processing. Although the source book is far from complete, it does provide a more extensive set of visual tasks that should conceivably be handled by a general purpose machine vision system. An incomplete listing of visual tasks is given here:

Visually guided manipulation (pick and place, assembly).

Visual inspection (surface, missing parts, relative dimension, sorting, labeling).

Visually guided navigation (road following, obstacle avoidance, landmark recognition, cross-country navigation).

Categorical perception (matching, classification, identification).

Expectation based on context, and context analysis.

Learning and memory; acquisition of models over time.

Synthesizing intermediate representation.

Manipulating internal models of complex, three-dimensional objects; geometric reasoning.

Matching the models against the data.

Selecting and representing models.

Selecting/matching model based on incomplete data.

Object recognition.

Shape from contour and surface.

Shape from motion, texture, lightness, color, depth.

Focus of attention; area of interest.

Selective attention.

Divided attention.

Segmentation.

Figure-ground discrimination.

Spatial relationships between shapes and parts of shapes.

Inside–outside relationship.

Gestalt groupings (good continuity, proximity, similarity, symmetry, common faith, closure).

Emergent feature configuration.

Visual counting.

Boundary tracking.

Rotation and scale invariance.

Illusory contours; boundary completion in absence of continuous data filling in the area within illusory contours.

Constancies (lightness, color, motion, scale).

From Visual Task to Neuronal Architecture

How can we judge that a selected visual task is fundamental to human visual performance? The answer requires the combination of knowledge from two subfields of neuroscience. We need to combine our knowledge of functional neuroanatomy and physiology of the visual system with clinical neuropsychology. The former provides bottom-up, detailed information about the anatomy and function of neural architectures underlying a particular visual task. The latter on the other hand, generates a top-down, global relationship between various aspects of the visual task and the coarse structures of the visual system.

All of these tasks at some lower level derive from primitive image attributes that include motion, texture, color, stereo, and perhaps edges and lightness. Specific information about depth, for example, can be extracted from a combination of stereo, motion, and lightness or shading. Motion in depth could be obtained from combining motion, edges, and stereo. Lightness, color, depth, and motion can tell us about light sources. In this way we should be able to enumerate all information at different levels that is necessary to support a high-level visual task such as obstacle avoidance. This exercise will also specify, in part, the connectivity between different modules. More significantly this approach might also eliminate the need for computationally expensive approaches such as representing visual functions as mathematically ill-posed problems (Poggio, 1985).

To illustrate the relationship between a visual task and its computational substrate, consider the visual task of Gestalt closure (see Fig. 2.3). We know from studying human vision that a visual deficit in illusory contours, which may be related to the pathology of area V2, is manifested by a reduced or limited performance in Gestalt closure tasks. Inability to do Gestalt closure leads to other failures in visual performance related to boundary finding. Because V2 is considered to be early on in the hierarchy of visual processes, this suggests that the Gestalt closure visual task seems fundamental to many other visual functions. The ability to perform tasks based on Gestalt closure translates into the development of a much more general machine vision system.

The end result of our preliminary research is a list of functions that underlie the performance of selected visual tasks. Table 2.2 presents neuropsychological results that aid us in relating functions and tasks to specific anatomical structures. The association of visual tasks with their possible neuronal substrate provides us with a methodology for synthesizing a framework for a general purpose machine vision system. Our map of the functional areas of the brain is a combination of the results of many studies (Hubel & Wiesel, 1963; Newsome, Maunsell, & Van Essen, 1986; Van Essen, Newsome, Maunsell, & Bixby, 1986). However, because the functional neuroanatomy of the Macaque monkey is not easily interpreted within the realm of computer science, we simplified this task by constructing functional diagrams like the one presented in Fig. 2.4. Interestingly, there are multiple hierarchies of processing stages, and the connectivity among various

TABLE 2.2
Visual Deficits and Corresponding Lesions

Visual Deficit	Symptom	Damage
Autopagnosia (Chusid, 1985)	Impaired recognition of body parts	Parietal lobe
Simultanagnosia (Roberts, 1984)	Only one aspect of an object can be appreciated at a time, e. g., color or shape	Left hemisphere or occipital lobe
Balint's syndrome (Roberts, 1984)	Inability to visually localize objects in space	Occipital lobe
Object mirror reversals (Feinberg & Jones, 1985)	Use common objects upside down or backward	Area 39—parietal-temporal-occipital association cortex
Visuoimaginal constructional apraxia (Grossi, Orsini, Modafferi, & Liotti, 1986)	Unable to draw a simple object without a model	Area 37—temporal posterior cortex
Charcot-Wilbrand syndrome (Chusid, 1985)	Unable to revisualize images	Posterior temporal cortex
Anton's syndrome (Chusid, 1985)	Agnosia where patients deny that they are blind	Area 7—parietal lobe
Gerstmann's syndrome (Chusid, 1985)	Disability to calculate, left-right disorientation	Area 39—parietal temporal association cortex
Prosopagnosia (Roberts, 1984; Damasio, Damasio, & van Hoesen, 1982; Benson, Segarra, & Albert, 1974)	Failure to recognize faces and complex objects	Bilaterial occipital association areas

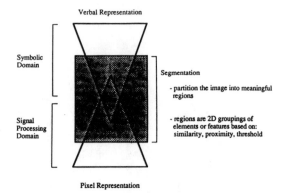

FIG. 2.4. The problem of segmenting images can be decomposed into top-down and bottom-up processes, which interact with each other at various depths depending on the task and its content.

modules, although nonrandom, allows for direct vertical and horizontal interactions between processing stages that are seemingly distant in function. This is unlike the clear, single-channel, sequential structures of general purpose vision proposed within computer vision where only immediately neighboring stages within the hierarchy can communicate with each other (Witkin & Tenenbaum, 1983).

ELEMENTS OF A GENERAL PURPOSE
VISION SYSTEM

As previous discussion shows, it is difficult if not impossible to completely specify general purpose vision at the present time. Hence, the remaining portion of this chapter is concerned only with selected elements of general purpose vision that have been less emphasized in computer vision. To simplify our consideration, we can make certain "commonsense" assumptions about a natural environment. For example, matter is cohesive; therefore, adjacent regions stick together in space and time. Most of the objects that we manipulate and interact with are solid. As such, one object can occupy only one point in space at one time. Many objects are symmetrical and they are usually attached to (stand on) some surface. In comparison to background, objects are small and surface properties are similar within the bounds of the object. To simplify the problem further, we can make some reasonable assumptions about a general purpose vision system. Multiple views of the object or a scene are available during the model formation (learning) stage.

Vision consists of various complex functions including recognition, model formation, visual task planning, and others. We emphasize visual recognition, which simplifies the problem by presupposing the existence of memories and world model. The human visual system is able to recognize objects of various positions/orientations and to describe highly complex scenes depicted in black-and-white photographs. This suggests that all of the information necessary for recognition can be encoded in only one variable—intensity changes. For computer vision, this is one of the most compelling justifications for using human vision as a model of general purpose vision systems.

Our overall philosophical approach (Skrzypek, in press) to the visual recognition problem is that the system consists of hierarchically arranged modules, interaction through feedforward and feedback pathways that can account for three different theories of shape recognition: feature detectors, template matching, and symbolic manipulation. Our approach thus separates the process of visual recognition into partially overlapping but distinct stages. For example, the lowest levels will always produce boundaries regardless of the higher processes that recognize boundaries as contours of the shape. Depending on the motivation, some of the boundaries may be considered to be irrelevant. It is our intent not to assume the availability of the input without specifying and generating the output from the preceding stage and to account for every processing step in terms of an explanation based on neural architectures.

Simple visual functions such as edge detection and textural segmentation (Mesrobian & Skrzypek, 1987; Skrzypek, 1987; Skrzypek & Mesrobian, 1992), color (Gungner, Skrzypek, & Heisey, 1992; Skrzypek, 1990b), illusory contours (Skrzypek & Ringer, 1992), and motion processing are carried out by the early stages. Complex pattern analysis, perceptual organization, short-term memory, attention, spatial perception, context analysis, and the formation of visual categories are carried out at the intermediate levels. The highest levels are concerned with generating goals, specifying tasks, planning their execution, long-term memory, multisensory interpretation, and so on. The whole system is loosely modeled on the multiple streams of information from the retina to higher centers (DeYoe & Van Essen, 1988; Livingstone & Hubel, 1988; Zeki & Shipp, 1988). The occipitotemporal stream, deals with identification of shape—what is the object? Another path, occipitoparietal, is concerned with spatial perception—where is the object?

Early Vision: Features and Spatial Frequencies

The complete set of processes at this level starts with retinal images that carry two-dimensional representation of intensity values encoding all information in the original scene. At the retinal level the information available includes (x, y) position, wavelength, luminance, local contrast, and motion and velocity. The retinal information flows to a higher center through parallel channels with distinct physiological and anatomical properties (Enroth-Cugell & Robson, 1966;

Rodieck, 1973). The retinal image attributes together with other functions computed at this early level (optical flow, stereo, motion, some aspects of textural segmentation, color [Heisey & Skrzypek, 1987]; lightness [Skrzypek, 1990a, Skrzypek & Gungner, 1992] and size invariances [Schwartz, 1980]; illusory contours [Skrzypek & Ringer, 1992]; etc.) are sufficient to describe surface properties such as luminance, hue, saturation, texture and perhaps shadows, highlights, and specular surfaces. This suffices to compute all other information about a scene, including two-dimensional attributes of simple features such as orientation, velocity, contrast, and disparity. Having surface properties and two-dimensional attributes of features, more sophisticated, three-dimensional abstractions can be computed such as shapes, spatial relationships, and three-dimensional movement. In the end, we see that a form can be computed from a combination of discontinuities in flow-fields, disparity, texture, color, shading, and luminance.

The specific connectivity between anatomically different "vision" modules has been the subject of concentrated research (DeYoe & Van Essen, 1988; Ungerleider & Mishkin, 1982; Zeki & Shipp, 1988). This multiple hierarchical structure of visual information allows for redundant, although maybe incomplete, representations of higher level attributes and necessitates some form of integration, perhaps by convergence.

Intracellular and extracellular recordings from single neurons at different points in the visual system suggest a sophisticated image feature analysis system (Hubel & Weisel, 1962; Ito, Fujita, Tanaka, & Cheng, 1992; Lettvin, Maturana, McCullogh, & Pitts, 1959; Neisser & Selfridge, 1960), including lines, monkey hands, color, depth, motion, bugs, sheep's faces, and so forth. Although sensory neurons are not feature specific and can be excited simultaneously by many stimulus attributes (wavelength, saturation, orientation, position, intensity, contrast, etc.), it is possible to extract unique representation of any feature, regardless of other attributes if proper sets of stimuli and receptive fields are considered.

The alternative to feature analysis theory is a hypothesis that the visual system analyzes spatial variations in image brightness. Here, receptive fields of neurons represent separate channels maximally tuned to different spatial frequencies (Campbell & Robson, 1968); neuron responses can be quantitatively analyzed without any reference to features. The best stimulus causes maximal response simply because it falls optimally within the sensitivity profile of a receptive field and the visual system acts as spatial frequency analyzer. This theory, derived from the studies of linear systems, assumes that the neural substrate underlying visual functions can be considered homogeneous, isotropic, and that a neural signal can be combined linearly. There are many other problems with the frequency analysis theory. For example, the four basic frequency channels include two that specialize in temporal information and two that specialize in spatial information. This is probably not enough to represent all spatial frequency with sufficient resolution. Interestingly, it has been suggested that all spatial frequencies are represented as a continuum among cortical neurons (De Valois, Albrecht, & Thorell, 1982) and in the lateral geniculate nucleus (Kaplan & Shapley, 1982). Also, there is no evidence that simple or complex cells can encode a phase information, an important component of the frequency domain. And the frequency bandwidth of most cortical neurons appears to be too wide to communicate certain spatial details of a shape. In addition, it is not clear how to synthesize amplitude and phase information within all bands of frequency into a coherent percept. Attempting to match frequency representation of a complete scene with data stored in memory might be extremely difficult. Finally, there are experimentally supported theories of perception, such as feature integration theory and texton theory, that cannot be easily explained in terms of spatial frequency analysis.

Higher Levels of Vision

The output of the earlier visual stages results in new representations in the form of perceptual primitives or their combinations that encode shapes or their parts. The abstraction of perceptual primitives out of image attributes involves Gestalt laws of grouping that might be preattentive, parallel mechanisms learned through interactions with environment. These mechanisms group the salient image attributes leading to emergence of a new configuration of existing or new shape features. Gestalt grouping is aided in turn by selective attention (Pomerantz, 1986).

The next several levels in our hierarchy can be loosely compared to the template theory (Trehub, 1977) of shape recognition. Simply, there are memories of prior retinal stimulation patterns that can be used as templates. The templates may be implemented as neural circuits, or they can be "learned" after a few presentations of a novel stimulus. Recent work by Goldman-Rakic (1992) suggested that Area 36 in the frontal lobe might serve as template locus for spatial vision. We expect that such short-term memories could also exist within such areas as V4, V5, and certainly IT or parietal lobe. Superimposing a memory (template) on the incoming pattern of activity generated by feature detectors of their Gestalt groupings would result in a match in the presence of the expected shape. We assume such problems as, for example, subjective contours, partial matches due to changes in distance, and so on, have been already addressed at the lower levels. For templates to be useful, they must be manipulated in three dimensions in order to account for missing surfaces of rotated three-dimensional objects that are projected onto two-dimensional retina. The information as to which direction to rotate can be precomputed from the response of feature detectors (Gungner & Skrzypek, 1986). Also, before templates are applied, some context information must be analyzed (Fuster, 1989), and the figure-ground segregation must be completed to some degree. These two functions are closely related and are in part responsible for the so-called image segmentation problem, which remains unsolved.

Finally, at the highest levels of the recognition hierarchy, we assume the existence of a symbolic scene representation that is the precursor of a verbal description. The AI community has successfully championed a localist symbolic approach for high-level visual reasoning and problem solving. Our symbolic representation differs from traditional AI by being distributed across the dynamic patterns of neural activity. In other words, visual information from the real world is modeled in the excitation patterns coming from widely distributed, lower level neural pathways. Using the convergent properties of self-organizing, adaptable neural architectures, symbols are distilled from the incoming patterned activity to produce a structural and functional description of the scene. The components of this symbolic representation come from lower levels and include categories of templates or short-term memories of perceived objects, expressed in a behavioral context. Long-term memory is the repository of distributed symbols measuring meaningful perceptions, each a stored representation of the activities corresponding to the patterned activity of lower modules. Operationally, the highest level uses symbols residing as prior memory patterns to modulate the activity of neural structures and to set up expectations of incoming data. Behavioral motivation regulates the combining of distributed memory primitives to assemble complex expectations and procedures in support of higher level visual tasks such as goal-directed scanning, sorting, and occluded object recognition. In this approach, perception is the byproduct of matching the current state of neural structures to an expectation (MacKay, 1967).

Control and Hypothesis Formation

Traditional engineering principles of organizing control of a mobile platform call for clean, modular decomposition of perception, task execution, motor control, reasoning, problem solving, planning, and others. Most of the realized or proposed computer vision systems are based on the idea of hierarchical organization of progressively more abstract functions where information progresses naturally (is abstracted) from lower level stages to higher levels of analysis, and each successive stage in the chain operates on all attributes generated by the preceding stage. This parallels similar concepts in the early days of natural vision (Hochberg, 1964; Hubel & Weisel, 1962).

It is conceivable that natural systems don't have such a nice and clean interface between various functions. Instead of one central representation of the world, there are, perhaps, many representations associated with various tasks. Evolutionary pressures organized these tasks, from most primitive to more sophisticated into one integrated system. Under some conditions, these task can complete with each other, whereas in other situations they may be synergistic. For example, at the reflex level, there is a direct connection between sensory input and motor output. With more sophisticated tasks, additional signal analysis and control can be added on top of this primitive system. Modular design of new subsystems on top of the existing ones would complicate the control but would also reduce the burden on the complexity of the local neural architectures. Hence, simple networks at higher levels could be made to influence the hierarchically lower networks leading to complex behavior. Feedback circuits are known to exist between prestriate and striate cortices and between thalamus and midbrain.

It is becoming clear that at least earlier stages of the visual system are composed of several channels processing input data in parallel. This accentuates the problem of controlling and integrating results of many specialized areas into one coherent percept. Each area of the cortex has multiple connections with other areas, both feedforward and back projections. Forward, lateral, and backward connections can be organized as both converging and diverging. Whereas convergence is useful for assembling and integrating information, the divergence underlies segregation, extrapolation. Multiple convergent and divergent connections between anatomically different visual areas suggest the architecture of many parallel hierarchies of functions. In general, when a function must be computed, any previously available result can be used.

In case of a goal-directed visual task, the processing within general purpose vision systems starts with the acquisition of the model of the expected object from the world knowledge base. Model-directed control refines an initial hypothesis and for some instance of it perhaps reconfigures the sensory system in order to find image-based evidence for the objects. Continuous refinement of hypotheses can be used to make predictions about the next data sample. As new sensory data is acquired, expectancies (current best guess) can be undated (Shneier, Lumia, & Kent, 1986). The expectancies derived from three-dimensional models can be used to generate predictions about two-dimensional projections of objects, which, in turn, predict the expected image features. Most of the "high-level" processes are couched in AI terminology because the neurosciences cannot at the present time specify the location or the architectures of neural structures underlying hypothesis formation or other aspects of model-driven control. Inferotemporal cortex cells, known to display longer response latencies, are perhaps involved in hypothesis formation (Fuster, 1989).

Because the system might consist of many knowledge sources that specialize in searching/interpreting objects and events in environment, we need a mechanism that allows the integration of their outputs into a coherent solution. Various forms of "winner-take-all" networks have been proposed. Symbol-based approaches to vision use blackboard mechanisms (Nagao & Matsuyama, 1980) to form hypotheses about potential objects and or relationships between them. In biological systems such notions might be based on feedback loops from a higher level to a lower level, although not all top-down pathways are well traced and understood. One conceivable principle of operation involves the existence of a feedback loop between function that is computed and the neural structure/architecture that computes this function. Expectation from higher levels act as templates that modulate feedback loops, thus filtering out only relevant information. In general, the problem of translating signals into symbols has not been solved.

Not all of the control issues pertain only to top-down information. Some control can be resolved locally by specifying rules of interactions between neuronal elements. Because the sensory system encodes relative information, the signaling is done as "above" or "below" an average level. For a system to avoid saturation requires the balancing of inhibitory (OFF) and excitatory (ON) channels (Kuffler, 1953; Skrzypek & Werblin, 1983; Skrzypek & Wu, in press). This dichotomy can also provide a phase information. The redundancy of information is also dictated by the inability of a neuron to generate negative spikes. However, this argument does not apply to graded potential cells. Other local control strategies include time-limited winner-take-all networks, automatic gain control via lateral inhibitory pathways, and various forms of an adaptive error detection and control.

Model Manipulation and Matching

Matching models with incoming and possibly incomplete image data entails the ability to manipulate object models. General purpose vision systems must possess mechanisms with which to maintain and manipulate hierarchical descriptions of object models. The object models could be represented using semantic networks with nodes representing objects and links representing constraint relations (Ballard, Brown, & Feldman, 1978). This approach is repeatedly used in many of the existing AI systems, and it has an intuitively obvious mapping onto neural networks. Although it is not clear whether semantic nets are the best representation for visual information, especially at the early and middle levels of processing, symbolic nodes could perhaps be implemented as small neural networks. Even in this case some of the nodes (neural nets) must be able to represent various aspects of objects in different contexts.

Matching models to data has been attempted in the past using rule-based systems. This is perhaps appropriate at higher levels of processing. Neural networks can be set up as a rule-based system where connectivity represents a prediction graph depicting expected objects or their representation in terms of some primitives. Incoming data then activates all neuronal feature detectors, but those that have been primed by the signal from the prediction graph generate highest level of activities. Highest activities in turn signify the correct matches. Thus matching is reduced to looking for the maximum cross section between activities in neuronal subgraphs representing image data and predicted object features. It is conceivable that some variation of this type of matching mechanism operates in a distributed fashion throughout all the levels of visual structure.

In other words, must of the matching is done perhaps on the local scale using feedback pathways and error detection implemented by networks of local, graded-potential neurons. This would mean that in our functional model there is no identifiable module responsible for matching but some abstracted results of all matches are recognized as compatible with expectancies derived from the long-term memory.

A distributed matching mechanism ("spreading activation" models) would help to avoid many difficulties encountered by traditional AI when attempting to solve the problems of recognition, hypothesis formation, goal seeking, belief maintenance, and so forth. In all of these problems, final interpretation of the scene depends on matching to establish correspondence between various representations. The goodness of a match depends on the number of features, labels, and attributes that are available for matching. Because we lack a strict definition of a shape, object, or a scene, exhaustive search of matching features can lead to combinatorial explosion. Hence, in traditional AI, to speed up the computation of correspondence, matching tends to be based on very few key features. In general, it has not been possible to enumerate all such features for all objects. However, using local matching within the distributed neural network, it could be possible to discover these features by abstracting from more primitive attributes. The solution to the category formation problem might be crucial in this task.

Adaptation, Learning, and Categorization

General purpose vision must be able to adapt its behavior through interactions with the environment; we cannot synthesize a system with a priori knowledge of all possible scenarios that it may encounter; perception cannot just simply be a filtering of incoming data but must be an active process of continuously matching incoming data to internally represented expectations. This implies that general purpose vision must be at times data driven to permit instant response in certain situations, whereas at other times it must be goal driven to permit the execution of requested tasks. ALVEN (Tsotsos, 1985) is an example of a system where the search through its knowledge based composed of object classes and relationships between them can be goal directed, model directed, and/or data directed. In other words, a general purpose vision system has to be able to select a prototype from input data, as well as descend down the categorical hierarchy under top-down control to instantiate the expected perceptual data. This implies that a general purpose vision system must be able to learn and categorize objects based on a limited number of presentations (Lynch, Grangner, Baudry, & Larson, 1988). Hence the need to analyze context, in parallel with, or even perhaps before, the final categorization of the input stimulus.

Consider the problem of maintaining correspondence between the data and its model while tracking/recognizing a moving object; besides the possible change in the appearance of an object due to rotation, translation, foreshortening, and so on, its background from one frame to another might also undergo a drastic change. For these reasons it seems desirable to have a match based on a few invariant, critical features. However, because we cannot enumerate all such critical features under all possible conditions, perhaps some form of adaptive learning is involved in selection of category prototypes.

Adaptation and learning, in addition to genetic encoding, account for a complexity and variety of behaviors as well as their adaptability to environmental changes. The key issue is how external stimuli are processed, analyzed, understood, abstracted, and transformed in context of previous knowledge, acquired skills, and motivations. Are there different kinds of memory and how many? How are they organized and stored in the brain? How large of a memory is needed to account, for example, for all visual experiences?

The processes by which stimuli are "encoded" into mnemonic representations remains unclear. Encoding, recorded as measurements of memory performance, was suggested to be influenced by semantic processing activities (Craik, 1979), by prior knowledge (DeGroot, 1985), by self-knowledge, and others. A critical aspect of remembering is the process of finding a cue that will retrieve the desired memory. It is not clear how retrieval cues are synthesized. Equally unclear is the process of how one knows that information is available in memory and whether knowing it is sufficient to recall it. A commonly understood notion of memory is that it allows a conscious and deliberate access to some very explicit knowledge about past events in their proper contextual setting. However, it is possible that there is more than one kind of memory.

Memory might involve various independent functions (limbic memory as declarative memory of events, places, and facts and nonlimbic, procedural memory of rules and algorithms for executing plans [Cohen & Squire, 1980]; a working memory as compared to the reference memory [Olton, Becker, & Handelmann, 1979]; or episodic as compared to nonlimbic being semantic) residing in different locations of the brain (Mishkin, 1979).

The original postulate introduced by Ramon y Cajal (1892), that learning leads to morphological changes in synaptic efficacy, was subsequently refined by Hebb (1949) into a specific proposal of cellular mechanism underlying memory and learning: (a) sensory stimulus could result in transient excitability change and enduring plastic change; (b) conjunction of activity in pre- and postsynaptic cells leads to strengthening of synaptic transmission.

Habituation and sensitization, two nonassociative foams of altering the behavior in response to a stimulus, have not been incorporated into a computer vision program. Classical conditioning is an associative form of learning; initially ineffective conditioned stimulus (CS) eventually generates a behavioral response, after it has been associated with some strong unconditioned stimulus (US). Classical conditioning is fundamental to learning about cause-and-effect relationships in the environment. It is conceivable that in all three of these behavior modification processes, some neural pathways for making associations are already available before the event. Classical conditioning, then, results in increased efficiency of preexisting pathways.

Most of the existing neural models of learning require an external teacher. These include Rumelhart et al.'s (1986) "Backpropagation," Widrow's "Adaline," (1960) and Fukushima's "Neocognitron" (1980). However, biological systems, especially during early stages of development, can perceive patterns (see section on Shape Perception) before they are capable of recognizing or communicating with a teacher. Hence, self-organization must be one of the principal rules of learning and adaptation (Adams, 1989; Carpenter & Grossberg, 1988; Linsker, 1986). Self-organization and self-regulation are well-recognized properties of the brain that have not been captured in any form by any existing machine vision system. The mechanisms underlying such ability derive from plasticity of the neural nets, which in absence of an external teacher must involve some internal controls such as adaptive error control, sensitization based on neighborhood activities, relative instead of absolute encoding, redundancy of identical information in various representations, and Hebbian rules of synaptic interactions. The flexibility of the cortex to develop feature detectors in response to incoming sensory information delivered to any specific part of the brain during some critical period has been confirmed by Sur and colleagues (Garrahty, Sur, & Roe, 1988).

In general, the problem of prototype ("basic level category") (Rosch & Loyd, 1978) selection from initial instances of data remains unsolved: Which critical features of shape or objects constitute prototypes? Another problem is that we don't know how they should be organized. In terms of neural structures, categorization is a process of adaptive selection of critical features or patterns of activities that are then stored in short-term memory. These patterns of activities are then recognized by matching them with expectancies. Familiar entities result in successful matches, whereas novel entities must be formed (via learning) into new categories of knowledge so that new experiences (i.e., new objects) can be retained for future use. Conceivably, patterns of activities generated by a hierarchy of feature detectors can activate dedicated neural structures (AIT or PIT) in a specific manner, which, in turn, selectively inhibit other neurons. This spatiotemporal pattern of activities could be at the basis of categorization of the inputs. The final result would be that a network learned to categorize input pattern by having a set of neurons that consistently display higher activity to this pattern.

Context Analysis

We assume that some part of the visual system is involved in continuous model building activities of the external world (Barlow, 1985). The activities of planning, task specification, and goal setting are part of these processes. Contrary to the traditional practice in the AI community, it is probably incorrect to arbitrarily divide the phenomenon of vision into high-level cognitive processes that can be investigated independently of early stages of visual processing. Meaningful perception at the higher level implies proper architectures and activities at the lower levels. Higher levels do not deal with the external world the way we see it, but they operate on the representation of this world derived from patterns of activities in some specific neural structures of earlier stages. And the modeling and interpretation that higher levels perform is realized in terms of manipulating activities and connectivity patterns of lower level neural nets. This defines contextual analysis of incoming images.

The visual system seems to incorporate a primitive but general function that permits the active investigation of a scene's topological organization before the analysis of details. When visual conditions are poor, figure-ground discrimination is easier to perform than the extraction of detailed shape information. This is consistent with the view that perception of details is easier in presence of context; context must be computed before the attention to details (Fuster, 1989). The context might include orientation of the figure, which, in turn, affects the perception of its three-dimensional surfaces. In tasks that depend on texture discrimination, local processes might be used to find surfaces. In tasks that depend on texture discrimination, local processes might be used to find discontinuities (Skrzypek & Mesrobian, 1992), but to complete the object's boundaries the system seems to utilize global configuration clues; detailed featural analysis of a shape takes place after top-down processes uncover the global configuration of a pattern. Why is this global structure so important? It seems that this information must be available to preattentive processes so that selective attention may be used in decisions, such as where to position the next saccade, when the system must attend to and at least coarsely analyze the image structure outside of the fovea in order to decide the future foveal fixation points. This implies that perceptual features are hierarchically organized and processed into objects beginning with figure-ground discrimination that perhaps drives selective attention. This is followed by an analysis of boundary information and finally details existing inside of the boundaries.

It is possible that some global analysis of the image, including perhaps figure-ground discrimination, is performed by the low-resolution, phylogenetically older, pulvinar-striate cortex path (Diamond & Hall, 1969). The geniculo-striate pathway, phylogenetically more recent, is dedicated to high-resolution analysis of the shape, once the target has been selected. In other words, it appears that perception of a large shape is partially independent of identity of its components and that it can be processed at different levels. Pomerantz (1983) proposed the existence of "emergent features" as the properties of the whole that may emerge from combination of primitive components but can nevertheless be detected directly just like hue and motion.

The question as to the order in which different processing levels should be initiated is less clear. The fact that context can enhance the perception of constituent parts argues against the strict bottom-up hierarchy of feature abstraction. It is conceivable that images extending into peripheral vision require bottom-up processing, whereas within foveal vision the top-down strategy reflects participation of a different attentional mechanism.

The context derives from correlates of expectations and resulting actions but it does not depend on information available to the previous fixations (Potter, 1975). The speed with which recognition of a scene takes place seems to depend on syntactic and semantic aspects of scene's coherence, which includes spatial relationships of objects, their size, and probability of occurrence in a given context (Biederman, 1981). This seems to argue against the bottom-up organization of an image; prior knowledge about spatial relationships of objects in a (familiar) scene is accessed in the process of image understanding. Similar ideas have been introduced in AI literature as a frame (Minsky, 1975) and schema (Hanson & Riseman, 1978). What remains unclear is how such a schema is invoked? It is possible that some "scene-emergent" features are perceived directly and, through access to some schematic representation of object relationships, trigger analysis of details. The alternative proposes that single features in the image are first identified, and this leads to creation of a prototype and eventually recognition of other objects.

Attention

The general purpose vision system should be able to perceive an object in environmental coordinates. This means that in some situations a general purpose vision system on a mobile platform must be able to discount image motion due to retina-motion or self-motion. Hence, a subsystem is needed that translates the information from retinal coordinates to egocentric coordinates and finally into environmental coordinates; retina-based visual field is small as compared to perception of space in environmental coordinates. The application of high resolution sensory array to a limited portion of a scene implies selective attention based on target motion or spatial position. These functions are computed by a dedicated module that attends to "extrapersonal" space. Selective attention based on the content of the stimulus (foveal attention) seems to be a more difficult problem that requires at least partial prior solution to shape recognition and context analysis.

At the highest levels, the cooperation between the task specifier, goal generator, and planner results in the motivationally selected design of routines necessary for the completion of a task. This "vision executive" (Davis & Kushner, 1987) might control the focus of attention mechanism for sensory data processing. The planner is also responsible for strategy selection, that is, the ordering of focus of attention rules and monitoring performance of the lower level processes, such as segmentation. The physiology of these very "high-level" functions of planing and specifying goals and tasks is not known except for the possibility that the frontal cortex might be involved.

One candidate for a top-down control process underlying attention operates by feedback pathways from higher levels where signal changes are gated by adjusting properties/shape of the receptive fields at lower level neurons (Wise & Desimone, 1988). A general principle of operation is that in a free running mode, the system always has some expectancy about the environment. This is one input to the mechanism that controls focus of attention. Any changes in the environment that are not predicted by the expected representation must be attended to. The attention mechanism can be focused on a small region of the environment without loosing data. Simply our world is highly structured, events are highly localized in time and therefore a few "well-chosen" (by adaptive learning) samples of the environment can provide all of the information necessary to maintain the match with the predicted model.

It is generally agreed that attention is a process of selection that allows the system to focus on one stimulus perhaps at the expense of the others. However it is not clear how it is decided what is relevant. Clearly, we can attend to specific attributes of a stimulus within one modality and we can select one modality over the others and can also pay attention to the stream of thoughts without any sensory input. For example, the visual system can shift attention from one point in visual space to another as demonstrated through saccadic eye movements, although we can also shift our attention without moving an eye. Conceivably, there are many subprocesses of attention related to various brain functions. The spatial attention mechanism, which seems to operate on peripheral vision, might consist of four subprocesses, including selection of a new target, disengaging current target, shifting attention to new target, and finally engaging a new target (Posner, 1984); patients who suffered parietal lesions could shift their attention and engage a new target, but it was difficult for them to disengage their attention from current target. Another mechanism could simply involve changing the resolution of the incoming representation (Crick, 1984; Treisman, 1988); not all of the scene needs to be viewed at all times.

How should a battery of high resolution sensors be directed to a target, and how is it decided what is important to look at? One strategy is to look only at changes in time and space. Static information is redundant. For this reason peripheral vision must be able to detect temporal changes and send information about spatial changes to processing centers that could foveate on the change. It is also important that attention be paid to the most critical events first. Peripheral motion in the frontoparallel plane is less critical then object motion toward the viewer. The top-down control should operate on a preselected model, using the current hypothesis about the scene to direct the next area to view. The strategy here is to initially perform coarse processing and, if warranted, follow it up with final analysis. Thus the foveation decision is in large part a function of bottom-up, low resolution information from peripheral vision that is supplemented by high-level information from the planning system in conjunction with short-term memory of previous foveations.

One critical problem in studies of attention that remains open is that context must be processed before attention to details. Conceivably the context is just a simple spatial representation in the parietal cortex that is continuously updated with information about past foveations and saccades. Thus, the attention mechanism, driven by input from either short- or long-term memory, specifies the next location to process, and only then are details available from the specific intrinsic maps. Integration of these features and their subsequent matching to some expectancies in long-term memory represents the final step of perception (recognition). Thus attention could be involved in the process of integration and linking image and perceptual attributes into parts of objects and into complete objects. Subsequent match of such representation to memory would lead to recognition (see Treisman & Gelade, 1980).

Shape Perception: Perceptual Organization and Segmentation

Object recognition is one of the fundamental tasks in perception, and it has been studied more than all other areas combined. For this reason, we limit our discussion here to a few cursory points. Clearly we lack an acceptable definition of an object and shape. Various proposals for representing objects are abundant (Ballard & Brown, 1982; Besl & Jain, 1985). Most representations developed within computer vision are not general enough to allow easy description of sculpted surfaces. One exception is the so-called surface boundary representation (SBR) (Ballard & Brown, 1982), which conveys the information about a three-dimensional surface in the form of triangle-faced polyhedrons. More complex techniques using quadrics and higher order polynomials have also been developed. It is conceivable that some variation of this representation would be well suited for neural network architecture.

The ability to recognize, classify, and identify objects from projections on the retina is a process that develops over time. It involves learning and structural changes to neural architectures. For example, 1-month-old infants prefer to look at grating patterns, and this preference changes after 2 months to bulls-eye patterns (Fantz & Nevis, 1976). With time, more and more complex patterns are preferred, which implies increasing ability to process details with higher spatial frequencies. The principles behind the development of mechanisms that underlie perceptual organization, that is, the organization into meaningful segments of an image, are not well understood, but it appears that most of the Gestalt principles are fully developed within 1 year after birth. For example, proximity (neighboring elements most likely represent the same surface) develops at 7 months of age, common fate (neighboring elements that move in the same direction belong to the same rigid body) is present at 1 month, subjective contour perception is detectable at 4 months (Campos, Bertenthal, & Haith, 1980), and symmetry (grouping elements that are symmetrical) is detectable at 5 months (Ferdinandsen, Fisher, & Bornstein, 1981). Similarly, the ability to perceive constant shape of varying size develops within the first year of life (Ruff, 1978). Furthermore, this implies that transformation from the viewer center coordinates to an object centered representation is a necessary prerequisite to discount object changes as the viewer moves around the environment.

Shape can be perceived at different levels of hierarchical decomposition. For example, we can attend to a shape of a car, or the door, or the mirror on the door, and so forth. The recognition appears to follow a match of a given level of detail with memory, and this process is serial in nature, from level to level (Hoffman, 1980), but it is not clear whether the visual system as a rule always attends to global aspects of the shape first.

Knowledge about the physiology of shape perception is limited to early stages of visual processing. Our knowledge becomes more general at higher levels. For example, we know in general that lesions of upper parts of V2 and V3 degrade pattern discrimination and visual acuity, whereas lesions in the lower part of these structures affect recognition of patterns and objects. However, we lack detailed knowledge of receptive fields and synaptic interactions at this level. We know that at the level of V4 and V5 retinotopic mapping is encoded in the cell response. All higher level areas are specialized for nontopographic processing of separate attributes of the image (Barlow, 1982). Some form of linking based on similarity of information in texture, color, motion, collinearity, disparity, figure/ground, and others must also take place in nontopographic representation. The underlying principles of linking are not well understood. Linking is important in the segmentation process because often, specific intrinsic images might not have enough information to be interpretable.

The fundamental problem of segmentation is that it cannot be considered as only a bottom-up process or as only a top-down process. Segmentation seems to involve both strategies, continuously interacting and penetrating each other to different depths, depending on the task. Can local information be sufficient for segmentation and interpretation? It is probably sufficient for partially guiding segmentation. To have the interpretation, we need also to include global information. The difficulty is that, except for trivial cases, it is not clear what makes up the global information and how to compute it. For example, segmentation based on the similarity of features within a region can be completed to a degree using only local information. However, it is not completely clear how similar the regions must be. Considering camouflage, similarity is not equivalent to identity. Hence, the question of which similarity parameters are most important in any specific situation might be guided by some global information.

Robust image segmentation will also reduce potential errors in higher level processes like planing and matching. Image segmentation can be based on cooperative/competitive relaxation algorithms (Hanson & Riseman, 1978) applied to all image attributes, as well as to integrated representations of intrinsic images. The segmentation process must take advantage of all available top-down strategies based on applicable knowledge. Some initial plans can be generated by bottom-up, coarse region segmentation. Such a plan could produce a set of large areas within each intrinsic image that become refined by integration with information from different image attributes. Eventually, a top-down process, initiated by the focus of attention module, segments the scene into background and target objects that can be examined for detailed structure in the context of large patches that might have already been interpreted as background.

Constancies and Generalization

Why are constancies important? All of the vision problems can be simplified if the system has the ability to deal with color, motion, shape, size, and lightness constancies. In some respect we can view constancy mechanisms as precursors of generalization. For example light and color constancy help to generalize across variations of illuminants. Size constancy allows generalization across varying shape sizes. Having constancy built into a neural network reduces the complexity of object recognition and minimizes the storage requirement. This implies that the process of categorization is simplified. It is conceivable that this form of generalization applies also to other attributes of shapes such as motion and texture. The concept of constancy also applies to shape but at a higher functional level. Thus we are able to generalize across the birds or vehicles despite their often drastic differences in details. It is conceivable that at this level the constancy might apply to spatial relationships among parts of the category members. In all of these cases the underlying concept is that constancy is a form of generalization that discounts the variation in the object caused by environmental changes.

The perception of surface lightness depends on its reflectance, position, and spectral characteristic of the direct illuminant, and the position, orientation, and reflectance of other surfaces that may contribute indirect, reflected light. Lightness constancy is the ability to separate these confounding components so that perception of a surface remains constant despite the variation in the nature of illumination. It seems intuitively clear how to implement this aspect of the lightness constancy (Skrzypek & Hoffman, 1989) and color constancy (Heisey & Skrzypek, 1987) with neural elements. The basic principle here is to enhance and compare variations that are above or below the average of the neighborhood by employing laterally interacting networks (Skrzypek, Mesrobian, & Gungner, 1989). This model, however, is inconsistent with perception based on binocular viewing of the scene. This implies that constancy might be computed after depth information has been made available. Furthermore, it implies that more global information about intensity discontinuities in three dimensions is necessary to separate components of illumination, surface orientation, and reflectance (Todd & Mingolla, 1983). This information cannot be computed locally; it requires the solution to global perceptual organization of an image.

Changes in size and sometimes orientation or even shape of the stimulus may cause little or no perceptual response. It is conceivable that size constancy is a real general function. We are capable of discounting the changes in retinal area that is covered by a familiar object when it moves in depth, and usually we can correctly use such information to judge the real size of an object. However, this capability ceases when there are no contextual cues about depth. Hence the neuron for size constancy must be at a higher level of processing, and it probably has a very large and complex receptive field that can integrate information from many other sources.

Clearly orientation constancy, if it exists, cannot be general because it is difficult for us to recognize otherwise familiar faces or characters if they are upside down. It seems intuitively clear how to implement some constancies with neural elements, for example, lightness (Skrzypek, 1990a) and color constancy (Heisey & Skrzypek, 1987). Shape constancy, however, is a difficult problem that might require cooperation among many complex distributed processes, including memory, in order to produce the final percept.

Color constancy can also be viewed as a multilevel phenomenon. In some sense it allows us to generalize the perception of the hue of a target despite spectral variation in the illuminant. On the local scale, it performs gain control and perhaps color contrast. Of course, like the lightness constancy, there are limits to color constancy. If there is no contextual information, such as hues of various background, or if the spectrum of illuminant is limited to one wavelength, only one hue will be perceived. It is known that color opponent cells can be found already in the retina. However, this mechanism cannot explain constancy in color naming.

CONCLUSIONS

The goal of developing a general purpose machine vision system has been decomposed into the series of subgoals. The analysis of existing computer vision systems elucidated not only the lack of truly robust machine vision, but also the need for a good working definition of a general purpose vision system. This analysis also demonstrated the need for collaboration between the neurosciences, computer science, and psychology. We have looked at the human visual system for hints regarding the underlying mechanisms necessary for the development of general purpose vision. Several conclusions can be drawn from this study. The definition and specification of general purpose vision systems could be made easier if a list of all visual tasks was created and, whenever possible, their corresponding neuronal substrates were noted. Although our list is far from complete, it serves to illustrate the tremendous scope of problems that a general purpose vision system must not only address but also solve.

Another conclusion is that, despite continuing great progress over the past 30 years, the problem of image segmentation remains unsolved: segmentation of images into well-defined regions that could be labeled as objects, surfaces, and so on. Although segmentation has been proposed as an intermediate stage of vision, buried somewhere in the bottom-up/top-down matching process, its functional position is not really clear, and it appears to be task and goal dependent. Segmentation is a process that operates on incoming data and is probably driven by prior knowledge or models that exist in the system. This is an area of high-level vision that we know little about. Perhaps a better description would be many processes at different levels in the hierarchy that cooperate/compete to produce a parceled image, where top-down control could be data-driven goals or tasks.

To illustrate the segmentation problem, consider vision as two triangles (see Fig. 2.4). The wide base of the bottom triangle represents image pixels (outputs of photoreceptors) conveying massively parallel input data. The narrowing arms of the triangle represent increasing abstraction and decreasing number of data. This hierarchy proceeds from numerical signals at the input to some intermediate form of symbolic representation. In other words, the phenomenon of vision begins in the signal processing domain; image capturing operations within the vertebrate retina or the electronic camera are best characterized in terms of signal processing. At the very top, on the other hand, we have a symbol processing domain where abstract object representations are converted into verbal description of a scene. In other words, the width of the upper, inverted triangle conveys the abstractness of a highly symbolic representation such as natural language and proceeds down to more physically grounded representations of the scene given, say, in the form of pattern activities in multiple neural networks.

It appears that segmentation consists of one or more processes that lie at the intersection of the triangles. Depending on the task, goal, or data, the segmentation processes may be few or many; that is, the upper and lower triangles can overlap to a greater or lesser extent. Segmentation is difficult because at some levels it is data driven, at some levels it is knowledge driven, and at other levels it is perhaps driven by both data and knowledge. Therefore, this transition from intensity data structures to symbolic data structures is ill-defined and very difficult—very difficult to understand in terms of natural systems, and very difficult to realize in terms of machine vision systems. A list of visual tasks and various functions corresponding to different positions of these two triangles would help with unraveling the problem of segmentation.

Starting with existing data from functional neuroanatomy, we synthesized a prototype framework for a general purpose vision system. This framework was then used to synthesize a more elaborate functional block diagram for a general purpose machine vision system (see Fig. 2.5). Two strongly shaded areas are emphasized because we know more about them from neurophysiology, and we also have some understanding of how to model their functionality in computer science. However, other areas are less well known. For example, categorization,

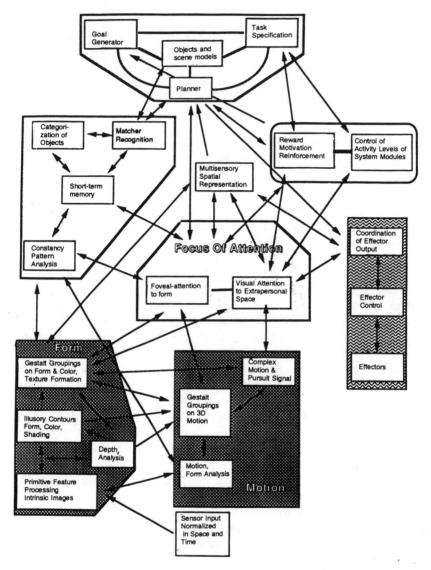

FIG. 2.5.   A decomposition of the visual process into functional areas related to specification of a general purpose machine vision.

constancy, matcher/recognition, and short-term memory are perhaps equivalent to visual areas like the posterior inferotemporal cortex (PIT) and the anterior inferotemporal cortex (AIT). A difficulty that remains in this attempt to synthesize a general purpose machine vision system is to determine where to place the homunculus that will finally decide what is being seen. Currently, our homunculus sits in the boxes labeled goal generator, task specification, and planner. These functions have received more attention in AI studies, and very little is known about their physiology. Some of their aspects may be analogous to parts of the frontal cortex and the limbic system. Our model is evolutionary and will improve with the collection of new experimental data from the simulations in conjunction with a continuous analysis of neuroscience literature.

Having a specification for a general purpose vision system will, we hope, permit us to develop successful evaluation methods that will perhaps be used as standards or guidelines for proposing new machine vision systems. It is clear that the development of such specifications will require collaborative efforts among neurophysiology, neuropsychology, computer science, and cognitive psychology.

ACKNOWLEDGMENTS

Support for the UCLA Machine Perception Laboratory environment is provided in part by generous grants from IBM and Hewlett Packard. We sincerely acknowledge support by ARCO-UCLA Grant 1, MICRO-Hughes grant 541122-57442, ONR grant N00014-14-86-K-0395, ARO grant DAAL03-88-K-0052, and PMTC-ATI grant N00123-87-D-0364. Special thanks to D. Gungner, E. Mesrobian, and other students of the MPL for critical reading of previous versions of this manuscript.

REFERENCES

Adams, J. L. (1989). *Principles of complementarity, cooperativity, and adaptive error control in pattern learning and recognition: A physiological neural network model tested by computer simulation.* Unpublished doctoral dissertation, University of California, Department of Computer Science, Los Angeles.

Ballard, D. H., & Brown, C. M. (1982). *Computer vision.* Englewood Cliffs, NJ: Prentice-Hall.

Ballard, D. H., Brown, C. M., & Feldman, J. A. (1978). An approach to knowledge-directed image analysis. In A. R. Hanson & E. M. Riseman (Eds.), *Computer vision systems* (pp. 271–282). New York: Academic Press.

Barlow, H. B. (1982). *General principles: The senses considered as physical instruments.* Cambridge: Cambridge University Press.

Barlow, H. B. (1985). *Cerebral cortex as a model builder.* New York: Wiley.

Baumgartner, G., von der Heydt, R., & Peterhans, E. (1984). Illusory contours and cortical neuron responses. *Science, 224,* 1260–1262.

Benson, D. F., Segarra, J., & Albert, M. (1974). Visual agnosia-prosopagnosia, a clinicopathological correlation. *Archives of Neurology, 30,* 307–310.

Besl, P. J., & Jain, R. C. (1985). Three-dimensional object recognition. *ACM Computing Surveys, 17,* 75–145.

Biederman, I. (1981). *On the semantics of a glance at a scene.* Hillsdale, NJ: Lawrence Erlbaum Associates.

Bolles, R. C., Howard, P., & Hannah, M. J. (1983). The 3dpo: A three-dimensional part orientation system. *Proceedings of the Ninth International Joint Conference on Artificial Intelligence*, pp. 1116–1120.

Brooks, R. A. (1981). Model-based computer vision. *Computer Science: AI., Vol. 14*, Ann Arbor, MI: UMI Resarch Press.

Campbell, F. W., & Robson, J. G. (1968). Application of fourier analysis to the visibility of gratings. *Journal of Physiology, 1975*, 551–566.

Campos, J. J., Bertenthal, B. I., & Haith, M. M. (1980). Development of visual organization: The perception of subjective contours. *Child Development, 51*, 1072–1080.

Carpenter, G. A., & Grossberg, S. (1988). The art of adaptive pattern recognition by self-organizing neural network. *Computer, IEEE, 21*, 77–88.

Chusid, J. G. (1985). *Correlative neuroanatomy and functional neurology* (19th ed.). Los Altos, CA: Lange Medical Publications.

Cohen, N. J., & Squire, L. R. (1980). Preserved learning and retention of pattern analysing skill in amnesia: Dissociation of "knowing what" and "knowing that." *Science, 210*, 207–209.

Craik, F.I.M. (1979). Human memory. *Annual Review of Psychology, 30*, 63–102.

Crick, F. (1984). The function of the thalamic reticular complex: The searchlight hypothesis. *Proceedings of the National Academy of Sciences, 81*, 4586–4590.

Damasio, A.R., Damasio, H., & van Hoesen, G. W. (1982). Prosopagnosia: Anatomic basis and behavioural mechanisms. *Neurology (NY), 32*, 331–341.

Davis, L. S., & Kushner, R. T. (1987). Vision-based navigation: A status report. *Proceedings of DARPA Image Understanding Workshop*, 153–169.

De Groot, A. D. (1985). *Thought and choice in chess*. New York: Basic Books.

De Valois, R. L., Albrecht, D. G., & Thorell, L. G. (1982). Spatial frequency selectivity of cells in macaque visual cortex. *Vision Research, 22*, 545–559.

DeYoe, E. A., & Van Essen, D. C. (1988). Concurrent processing streams in monkey visual cortex. *Trends in Neurosciences, 11*(5), 219–226.

Diamond, I. T., & Hall, W. C. (1969). Evolution of neocortex. *Science, 164*, 251–262.

Ekstrom, R. B., French, J. W., Harman, H. H., & Dermen, D. (1976). *Kit of factor-referenced cognitive tests*. Princeton, NJ: Educational Testing Service.

Enroth-Cugell, C., & Robson, J. G. (1966). The contrast sensitivity of retinal ganglion cells of the cat. *Journal of Physiology (London), 187*, 517–552.

Fantz, R. L., & Nevis, S. (1976). Pattern preferences and perceptual-cognitive development in early infancy. *Merril-Palmer Quarterly, 13*, 77–108.

Feinberg, T. & Jones, G. (1985). Object reversals after parietal lobe infarction. *Cortex*, 261–271.

Ferdinandsen, K., Fisher, C. B., & Bornstein, M. H. (1981). The role of symmetry in infant form discrimination. *Child Development, 52*, 457–462.

Fodor, J. A. (1983). *The modularity of mind*. Cambridge, MA: MIT Press.

Fukashima, K. (1980). Neocognition: A self-organizing neural network model for a mechanism of pattern recognition unaffected by shift in position. *Journal of Biocybernetics, 36*, 193–202.

Fuster, J. M. (1989). Inferotemporal units in selective visual attention and short term memory. (Submitted.)

Garrahty, P. E., Sur, M., & Roe, A. W. (1988). Experimentally induced visual projections into auditory thalamus and cortex. *Science, 242*, 1437–1441.

Gotlieb, C. C. (1985). *The economics of computers: Cost, benefits and strategies*. New York: Prentice-Hall.

Grossi, D., Orsini, A., Modafferi, A., & Liotti, M. (1986). Visuoimaginal constructional apraxia: On a case of selective deficit of imagery. *Brain and Cognition, 5*, 255–267.

Gungner, D., & Skrzypek, J. (1986). A connectionist architecture for matching 3-D models to moving edge features. *Proceedings of SPIE Conference on Intelligent Robots and Computer Vision, 726*.

Gungner, D., Skrzypek, J., & Heisey, I. (1992, January). *Double opponency mechanism and color constancy* (Tech. Rep. UCLA-MPL-TR 92-5). Los Angeles: University of California, Machine Perception Laboratory.

Hanson, A. R., & Riseman, E. M. (1978). Visions: A computer system for interpreting scenes. In A. R. Hanson & E. M. Riseman (Eds.), *Computer vision systems* (pp. 303–334). New York: Academic Press.

Harmon, L. D., & Lewis, E. R. (1966). Neural modeling. *Physiological Review, 46*, 513–591.

Hebb, D. O. (1949). *The organization of behavior.* New York: Wiley.

Heisey, I., & Skrzypek, J. (1987). Color constancy and early vision: A connectionist model. *Proceedings of the IEEE First Annual International Conference on Neural Networks, 4*, 317–325.

Herman, M., & Kanade, T. (1984). The 3d mosaic scene understanding system: Incremental reconstruction of 3d scenes from complex images. *Proceedings of DARPA Image Understanding Workshop,* pp. 137–148.

Hochberg, J. E. (1964). *Perception.* Englewood Cliffs, NJ: Prentice-Hall.

Hochberg, J. (1987). Machines should not see as people do, but must know how people see. *Computer Vision, Graphics, and Image Processing, 37,* 221–237.

Hoffman, J. E. (1980). Interaction between global and local levels of a form. *Journal of Experimental Psychology: Human Perception and Performance, 6,* 222–234.

Hubel, D. H., & Wiesel, T. N. (1962). Receptive fields, binocular interaction and functional architecture in the cat's cortex. *Journal of Physiology (London 160),* 106–154.

Hubel, D. H., & Wiesel, T. N. (1963). Shape and arrangement of columns in the cat's striate cortex. *Journal of Physiology (London), 165*(3), 559–568.

Ikeuchi, K. (1981). Recognition of 3d objects using the extended gaussian image. *Proceedings of the Seventh International Joint Conference on Artificial Intelligence,* 24–28.

Ito, M., Fujita, I., Tanaka, K., & Cheng, K. (1992, November). Columns of visual features of objects in monkey inferotemporal cortex. *Nature, 360*(6402), 343–346.

Kaplan, E., & Shapley, R. M. (1982). X and y cells in lateral geniculate nucleus of macaque monkey. *Journal of Physiology, 330,* 125–142.

Kuffler, S. W. (1953). Discharge patterns and functional organization of mammalian retina. *Journal of Neurophysiology, 16,* 37–68.

Lettvin, J. Y., Maturana, H. R., McCullogh, W. S., & Pitts, W. H. (1959). What the frog's eye tells the frog's brain. *Proceedings of the Institute of Radio Engineers, 47,* 1940–1951.

Linsker, R. (1986). From basic network principles to neural architectures: Emergence of spatial-opponent cells. *Proceedings of the National Academy of Sciences, USA, 83,* 7508–7512.

Livingstone, M., & Hubel, D. (1988, May). Segregation of form, color, movement, and depth: Anatomy, physiology, and perception. *Science, 240,* 740–749.

Lynch, G., Grangner, R., Baudry, M., & Larson, J. (1988). *Cortical encoding of memory: Hypotheses derived from analysis and simulation of physiological learning rules and anatomical structures.* Cambridge, MA: MIT Press.

MacKay, D. M. (1967). *Ways of looking at perception.* Boston, MA: MIT Press.

Marr, D. (1976). Prosopagnosia: Anatomic basis and behavioral mechanisms. *Vision Research, 32,* 331–341.

Marr, D. (1982). *Vision.* Los Altos, CA: Lange Medical Publications.

McKeown, D. M., Harvey, W. A., & McDermott, J. (1985). Rule-based interpretation of serial imagery. *IEEE Transactions on Pattern Analysis and Machine Intelligence, 7*(5), 570–585.

McLelland, J. L., Rumelhart, D. E., and the PDP Research Group. (1986). *Parallel distributed processing.* Cambridge, MA: MIT Press.

Mesrobian, E., & Skrzypek, J. (1987). A connectionist architecture for computing textural segmentation. *Proceedings of the SPIE Conference on Image Understanding and the Man-Machine Interface,* pp. 48–52.

Minsky, M. (1975). *A framework for representing knowledge.* New York: McGraw-Hill.

Mishkin, M. (1979). Analogous neural models for tactual and visual learning. *Neuropsychologica, 17*, 139–151.

Nagao, M., & Matsuyama, T. (1980). *Structural analysis of complex aerial photographs.* New York: Plenum.

Nazif, A. M., & Levine, M. D. (1984). Low-level image segmentation: An expert system. *IEEE Transactions on Pattern Analysis and Machine Intelligence, 6*(5), 555–557.

Neisser, U., & Selfridge, O. G. (1960). Pattern recognition by machine. *Scientific American, 203*, 60–68.

Nevatia, R., & Price, K. E. (1982). Locating structures in aerial images. *IEEE Transactions on Pattern Analysis and Machine Intelligence, 4*(5), 476–485.

Newsome, W. T., Maunsell, J.H.R., & Van Essen, D. C. (1986). Ventral posterior visual area of the macaque: Visual topography and area boundaries. *Journal of Comparative Neurology, 252*, 139–153.

Ohta, Y. (1980). *A region-oriented image analysis system by computer.* Unpublished doctoral dissertation, Kyota University, Department of Information Science.

Olton, D. S., Becker, J. T., & Handelmann, G. E. (1979). Hippocampus, space and memory. *Behavioral and Brain Science, 2*, 313–365.

Poggio, T. (1985). Early vision: From computational structure to algorithms and parallel hardware. *Comp. Vis. Graph. Imag. Proc., 31*, 139–155.

Pomerantz, J. (1983). Global and local precedence: Selective attention in form and motion perception. *Journal of Experimental Psychology, 112*, 516–540.

Pomerantz, J. R. (1986). *Visual form perception: An overview.* New York: Academic Press.

Posner, M. I. (1984). *Neural control of the direction of covert visual orienting* (Tech. Rep. No. TR 84-4). Portland: University of Oregon.

Potter, M. C. (1975). Meaning in visual search. *Science, 187*, 965–966.

Ramon y Cajal, S. (1973). La Retine des Vertebres. *The vertebrate retina: Principles of structure and function* (pp. 775–904). San Francisco: W. H. Freeman.

Roberts, J. K. (1984). *Differential diagnosis in neuropsychiatry.* New York: Wiley.

Rodieck, R. W. (1973). *The vertebrate retina, principles of structure and function.* San Francisco: W. H. Freeman.

Rosch, E., & Loyd, B. (1978). *Cognition and categorization.* Hillsdale, NJ: Lawrence Erlbaum Associates.

Rubin, S. M. (1978). *The argos image understanding system* (Tech. Rep. NTIS Order No. AD-A066736). National Technical Information Service.

Ruff, H. A. (1978). Infant recognition of invariant form of objects. *Child Development, 49*, 293–306.

Schwartz, E. L. (1980). Computational anatomy and functional architecture of striate cortex: A spatial mapping approach to perceptual coding. *Vision Research, 20*, 645–669.

Shirai, Y. (1978). Recognition of real-world objects using edge cues. In A. R. Hanson & E. M. Riseman (Eds.), *Computer vision systems* (pp. 353–362). New York: Academic Press.

Shneier, M. O., Lumia, R., & Kent, E. W. (1986). Model-based strategies for high-level robot vision. *Computer vision, graphics, and image processing, 33*, 293–306.

Skrzypek, J. (1987, July). Sensing and perception: Connectionist approaches to subcognitive computing. *Proceedings of the NASA Conference on Space Telerobotics.*

Skrzypek, J. (1990a). Lightness constancy: Connectionist architecture for controlling sensitivity. *IEEE Transactions on Systems, Man, and Cybernetics, 20*(5), 957–968.

Skrzypek, J. (1990b). Neural specifications of a general purpose vision system. *Proceedings of ACNN'90, Australian Conference on Neural Networks,* 160.

Skrzypek, J. (in press). Neural specification of a general purpose vision system. In O. Omidvar (Ed.), *Progress in neural networks* (Vol. 2). Norwood, NJ: Ablex.

Skrzypek, J., & Gungner, D. (1992). Lightness constancy from luminance contrast. *International Journal of Pattern Recognition and Artificial Intelligence, 6*(1), 1–36.

Skrzypek, J., & Hoffman, J. (1989). *Visual recognition of script characters: Neural network architectures* (Tech. Rep. No. MPL 90-10). Los Angeles: University of California, Machine Perception Laboratory.

Skrzypek, J., & Mesrobian, E. (1992). Segmenting textures using cells with adaptive receptive fields. *Spatial Vision.*

Skrzypek, J., Mesrobian, E., & Gungner, D. (1989). Neural networks for computer vision: A framework for specification of a general purpose vision system. *Proceedings of SPIE Conference on Image Understanding and the Man-Machine Interface,* 1076.

Skrzypek, J., & Ringer, B. (1992). Neural network models for illusory contours perception. In *Proceedings of IEEE Computer Vision and Pattern Recognition* (pp. 681–683). New York: IEEE Press.

Skrzypek, J., & Werblin, F. S. (1983). Lateral interactions in absence of feedback to cones. *Journal of Neurophysiology, 49*(3), 1007–1016.

Skrzypek, J., & Wu, G. (in press). Computational model of invariant contrast adaptation in the outer plexiform layer of the vertebrate retina. In F. Eeckman (Ed.), *Computational neural systems CNS*92.* Kluwer Academic.

Todd, J. T., & Mingolla, E. (1983). Perception of surface curvature and direction of illumination from patters of shading. *Journal of Experimental Psychology: Human Perception and Performance, 8,* 583–595.

Torre, V., & Poggio, T. (1984, August). *On edge detection* (Tech. Rep. No. AI MEMO 768). Cambridge, MA: MIT Press.

Trehub, A. (1977). Neuronal models for cognitive processes: Networks for learning, perception and imagination. *Journal of Theoretical Biology, 65,* 141–161.

Treisman, A. (1988). Features and objects: The fourteenth Bartlett Memorial Lecture. *Quarterly Journal of Experimental Psychology: Human Experimental Psychology, 40,* 201–237.

Treisman, A. M., & Gelade, G. (1980). A feature integration theory of attention. *Cognitive Psychology, 12,* 97–136.

Tsotsos, J. K. (1985). Knowledge organization and its role in representation and interpretation for time-varying data: The alven system. *Computational Intelligence, 1,* 16–32.

Ungerleider, L. G., & Mishkin, M. (1982). *Two cortical visual systems.* Cambridge, MA: MIT Press.

Van Essen, D. C., Newsome, W. T., Maunsell, J.H.R., & Bixby, J. L. (1986). The projections from striate cortex (V1) to areas V2 and V3 in the macaque monkey: Asymmetries, area boundaries, and patchy connections. *Journal of Comparative Neurology, 244,* 451–480.

Wertheimer, M. (1938). Untersuchungen zur lehre von der gestalt. *Psychologische Forschung, 4,* 301–350.

Widrow, B., & Hoff, M. E. (1960). Adaptive switching circuits. *1960 WESCON Convention Record,* Part IV, pp. 96–104.

Wise, S. P., & Desimone, R. (1988). Behavioral neurophysiology: Insights into seeing and grasping. *Science, 242,* 739–739.

Witkin, A. P., & Tenenbaum, J. M. (1983). On the role of structure in vision. In B. Hope, J. Beck, & A. Rozenfeld (Eds.), *Human and machine vision* (pp. 481–544). New York: Academic Press.

Zeki, S., & Shipp, S. (1988, September). The functional logic of cortical connections. *Nature, 335,* 311–317.

3 Human Benchmarking of Natural Language Systems

Eva L. Baker
CRESST/University of California, Los Angeles

INTRODUCTION

This chapter documents an approach to the evaluation of intelligent computer systems, in particular, natural language (NL) understanding systems. Our overall project strategy was to develop a multidimensional system to evaluate both qualitative and quantitative elements of natural language computer programs. The reasons for this research are threefold. First, it is difficult for program managers in government and potential users of intelligent computer systems to get clear and consistent indicators of improvement in system performance in other than very technical terms. Second, the evaluation of such systems in the computer science community had, to this point, proceeded unsystematically and in general without regard to the long history in evaluation and measurement shared by the social sciences. Third, as a research enterprise, we were interested in understanding how and how much of computer programs purported to model intelligence (i.e., artificial intelligence) can be referenced back to the performance of humans.

Artificial intelligence (AI) systems that understand natural language offer a great advantage to users who are not trained in computer languages. Users can ideally interact with these AI systems using the language they would use when interacting with people. There are multiple applications in the natural language area. One such system that processes natural language is known as an interface, or front-end to a data base. An interface contains language-specific semantic and syntactic information that allows a user access to a data base through a natural language, usually by asking questions.

This chapter focuses on research in relating human performance measures to natural language implementations involving a front-end to a data base. Although our work extends to other areas, that is, vision (see Skrzypek, chap. 2) and expert systems (see O'Neil, Baker, Ni, Jacoby, & Swigger, chap. 1), we believe we had the best chance of success in the natural language area for two reasons. First, the natural language area is one of the most well developed in the AI community, with literally scores of programs in language understanding (DARPA, 1989; Jacobs, 1990; Lehnert & Sundheim, 1991). Second, from the educational measurement side, there is a long history and extensive set of testing approaches related to language and reading comprehension. Our task was essentially to determine if there were unions between these two traditions that would help us describe and assess natural language implementations in terms of measured human language proficiencies. We call our methodological approach *human benchmarking*.

HUMAN BENCHMARKING

Human benchmarking in the present context means quite simply comparing the performances of intelligent computer systems to the performances of humans on the same task—a variation of the Turing test, an attempt to create an interval scale. Whereas intelligent computer systems can clearly be evaluated on various levels, including efficiency and effectiveness at accomplishing the task, our focus was on the output of a system irrespective of the processes it might go through to produce that output. Clearly, this approach glosses over some important differences between systems and people. For example, we decided not (and it was not really feasible) to measure explicitly important other language performances that the human comparison group can accomplish in addition to those targeted by the system. People are obviously infinitely more creative and proficient in language than any existing system. The remainder of the chapter provides specific details of one of our studies developing human performance benchmarks for a specific natural language system.

The Natural Language System IRUS

The particular program used in the research is a natural language query system. Essentially, such a program permits the user to ask questions in regular English prose to another computer program, for instance, a data base or an expert system. This natural language system, IRUS (Bates, Stallard, & Moser, 1986), is an interface between the user and the set of desired information, and provides a rapid, natural, and convenient method for obtaining information. IRUS has been designed to serve as a general purpose interface to a broad range of data bases and expert systems. It is designed as a basic syntactic shell that needs to be filled with specific semantics in order to work. To be used, IRUS must be specifically adapted to each of its partner data bases or expert systems. The particular semantics (content) of the data base or expert system must be translated into rules used

TABLE 3.1
IRUS in Two Domains

IRUS in a library science domain:

> "Of the books on artificial intelligence, how many have been classified as textbooks?"
> "Have there been as many requests for books about medicine this year as we planned for in our budget?"
> "Which organizations that we receive reports from have responded to either of our recent questionnaires?"

IRUS in the domain of navy ships:

> "List the number of ships that are deployed in the Indian Ocean."
> "What's the name of the commander of Frederick?"
> "What is Vinson's current course?"

---

by IRUS. Table 3.1 shows a sample of queries that IRUS can deal with, in two different domains.

Our knowledge of what IRUS can do is based on a system test of IRUS, where the system successfully answered a series of 165 questions. We have taken these questions, classified them into semantic and syntactic categories, and developed a set of test specifications designed to measure human ability to understand these very same types of questions.

Because IRUS is always embedded in the specific domain of a data base, for example, a data base of navy ships in the Pacific Ocean, our measurement approach needed to separate out the understanding of the question from the ability to provide the answer. Clearly, making comparisons between the program and people based on the correctness of answers about the location of navy ships makes no sense, for a computer can have vastly more detailed information than any person. Because we believed that many of the IRUS query types could be answered by very young children, we developed a methodology that would provide for children a very simple data base—one consisting of animals, people, houses, their attributes, and positions. One goal was to connect IRUS question types to the simple data base and measure children's results. We recognize the controversy inherent in separating syntax from semantics, but the strategy seemed inviting in the light of feasibility.

Our methodology included a pretest that determines whether students understood the elements in the data base. Figure 3.1 presents an example of an item from the pretest. By screening out students who cannot identify the data base elements, we were able to infer that students' selection of the correct answer was based on their understanding of the question. In our study, unlike the real IRUS applications, the data base functioned only to permit us to assess language function. A sample performance test item is provided in Fig. 3.2. All tests were implemented in HyperCard and administered on Macintosh SE computers.

FIG. 3.1. IRUS vocabulary pretest. Students are shown a prompt (e.g., "Point to the snake") and asked to respond.

PILOT STUDY

In order to determine the appropriate language understanding level for students at which to administer the IRUS test, we piloted the IRUS test with early elementary school and preschool students. Depending on the type of query, students answered either with an oral response or by pointing to the answer on the computer screen. When the student responded orally, the administrator typed the answer so that it was entered in the computer. (For example, in Fig. 3.2 a student might answer the query "How many cats in the truck have striped balloons" by saying "one." The administrator would type "one" into the computer.) When the student pointed to an answer on the screen, the administrator used the mouse to highlight the student's choice. (For example, a student might answer the query "Choose the cats with striped balloons" by pointing to any cats on the screen that fulfilled the requirement. The administrator would click on each cat identified by the student.)

In a think-aloud procedure designed to validate students' understanding of the questions they had answered, students were asked to explain their responses to the more complex queries. After their response was entered in the computer, the administrator would ask, "Why did you say _____ " or "How did you know that _____ was the answer?" These responses were tape-recorded for analysis in conjunction with the test transcript produced by the Macintosh.

The results of the pilot study indicated the following:

> **How many cats in the truck have
> striped balloons?**

Answer = 1

Comparable IRUS queries:

How many ships in the Third Fleet are C–3?
How many of the ships in the Indian Ocean are C–5?

FIG. 3.2.   Test item.

1. Students reading at or above a second-grade level generally could recognize all of the elements in the data base when those elements were presented in the pretest.

2. Students reading at or above a second-grade level had no difficulty in reading aloud the test queries. Some students skipped over words and even paraphrased some words when reading aloud (for example, one student substituted the word "bike" for "bicycle").

3. Some students reading at or above a second-grade level preferred to process the query silently, rather than read it aloud, even when reminded that the test administration procedures called for them to read aloud.

4. Students reading at or above a second-grade level had difficulty with those queries that were more than nine words long and that contained more than one delimiter, when the relationship between delimiters were expressed by either the conjunction "and" or the conjunction "or."

5. Students reading at or above a second-grade level often answered a query with a response that was literally incorrect but pragmatically valid. For example, when asked "How many cars have striped flags?", some students responded by pointing to all the cars that met the stipulation rather than answering with a specific number.

6. Students reading at or above a second-grade level provided very little additional information when asked to explain their answers to test queries. Typical answers to the question "How did you know that was the correct answer?" were to point to specific details in the picture or to state "because that's the answer."

THE MAIN STUDY

Based on the pilot study, a final version of the methodology system was developed and used in the main phase of testing. This phase of the study represented the human benchmarking of IRUS. Empirical data were collected in the following manner.

Subjects

Three kindergarten classes and three first-grade classes from an elementary school located in southern California participated. A total of 151 students were tested; however, students who were enrolled in a concurrent English-as-a-second-language class and students who did not have a complete data set were excluded from the sample. In total, 126 students, 60 kindergartners and 66 first-graders, were included in analyses.

Instruments

Two tests were administered to these students: a revision of the pilot Natural Language Evaluation Test (NLET) based on IRUS queries (Baker, Turner, & Butler, 1990); and the language section of the Iowa Tests of Basic Skills (ITBS), which was used as a criterion measure to allow the grouping of students by national grade-equivalent norms. A discussion of each test follows.

NLET. This test was developed to explore the potential value of establishing human performance standards by which to measure the natural language understanding of an AI system. NLET questions were designed to duplicate the queries that a specific AI interface, IRUS, was able to understand within a domain appropriate for the test population. The NLET is a computer-delivered language test for children, the content of which reflects linguistic structures that IRUS can process. As in the pilot test, the NLET was programmed in HyperCard and administered individually to students on a Macintosh computer. For most students, test administration time was between 15 min and 18 min. The individual being tested saw illustrations on the computer screen and the administrator asked scripted questions about the illustrations. No examinee was allowed to manipulate the computer. The administrator entered each student response in the computer so that a transcript of the responses could be generated for scoring at a later time.

The specifications for the NLET version in this study consisted of nine sentence types based on various combinations of syntactic constituents. Table 3.2 displays the nine sentence types. It should be noted that, although the examples were given in their question forms where relevant, the queries were assigned to types based on their constituent structure before question transformation.

Table 3.3 displays NLET items for each of the nine sentence types with a corresponding IRUS query. Although the NLET items reflected the linguistic structure of the IRUS queries, the vocabulary of the NLET items was appropriate for children. The correct responses for the NLET items were based on how IRUS processed parallel queries, that is, queries with structures similar to the NLET items.

This NLET form contained a total of 39 test items. The items allowed for the measurement of external features defined by the specifications: pronoun and deictic reference, possession, ellipsis, and change in status over time.

The distribution of items across the specifications was not equal because pilot test results revealed some types of items to be much easier than others. It was not considered necessary to conduct extensive retesting of types that were extremely easy, although easy items were included to create opportunities for testing features such as pronoun reference and ellipsis.

A total reliability estimate for the NLET was not established because test construction (items in sets that are dependent on preceding items) prevents the use of any estimate of reliability based on internal consistency.

TABLE 3.2
NLET Test Specification Types

Specification Type	Constituents
1	Noun Phrase (NP) + Verb + NP
2	Performance Verb + NP
3	NP + Copula + NP
4	4NP + Verb or Copula + Adjective Phrase
5	NP + Verb or Copula + Adverb/Adverb Question
6	NP + Verb or Copula + Prepositional Phrase
7	NP + Verb + Relative Clause
8	Performance Verb + NP + Relative Clause
9	Nonreferential "there" + Copula + NP

TABLE 3.3
Example NLET Items and Parallel IRUS Queries by Item Type

Specification Type	NLET Item	IRUS Query
1	Does the dog have a hat?	Does FREDERICK have TASCOM?
2	Point to the car with a flag.	Display all carriers in the PACFLT.
3	Who is the driver of the truck?	What's HAMMOND's readiness?
4	Is he striped?	What ships ae TASM capable?
5	Why did he lose his balloon?	When was the last update to Midway's readiness?
6	How many striped snakes are on the roof?	How many cruisers are in WESTPAC?
7	How many of them [the cats that are in the truck] have balloons?	How many cruisers are deployed that are C3?
8	Point to the cars that have a flag or that have stripes.	List the ships that are C4 or that are C5.
9	Are there any snakes in the house that are striped?	Are there any submarines in the South China Sea?

The ITBS Language Section.   The need for a standardized test of language ability to reflect developmental differences in examinees' performances led to the selection of the Iowa Tests of Basic Skills (ITBS) for use in this project. As noted in the teacher's guide that accompanies the ITBS (Hieronymus, Hoover, & Lindquist, 1986), the test battery "is intended to assess the development of early educational experiences in the basic skills: listening, pre-reading, vocabulary, language and mathematics" and is "a continuous assessment program for Grades K–9" (p. 1). The continuous sequence of assessment offered by the ITBS allowed for a common measure of language proficiency across age groups.

Instead of the complete ITBS battery, the language section, which focuses on the measurement of students' "comprehension of linguistic relationships," was selected for inclusion in the study because its content was judged to be closer to the content of the NLET. Level 5 was administered to the kindergartners and Level 6 was administered to the first-graders. The ITBS *Manual for School Administrators* (Hieronymus & Hoover, 1986) reports reliability estimates of .77 for Level 5 and .81 for Level 6 of the language section.

Both the Level 5 language section and the Level 6 language section have 11 multiple-choice practice items and 29 multiple-choice test items. The ITBS guide suggests approximate administration times of 10 min for the practice items and 20 min for the 29 test items. However, the ITBS is not a timed test and "time requirements are expected to vary with different examiners and groups."

The first-graders were able to complete all 40 items in one 45-min session. However, two sessions were required for the kindergartners to complete all the items. Extra time was required because of their lack of familiarity with the task, limited concentration, and the availability of only one proctor in each classroom. The kindergartners completed the practice questions and the first 10 test questions during one 30-min sitting and the remaining 19 questions during a second sitting of 25 min, scheduled within 3 days of the first session. The same test administrator conducted all of the ITBS testing.

RESULTS

Descriptive statistics for the Level 5 ITBS language section (kindergartners), the Level 6 ITBS language section (first-graders), and the NLET are reported in Table 3.4. For the NLET, the mean and standard deviation for the combined set of kindergartners and first-graders is also reported. A mean and standard deviation for kindergartners plus first-graders on the ITBS cannot be reported because the students were not administered the same form of the test.

The descriptive statistics indicate that the two groups, kindergartners and first-graders, performed similarly on both tests used in this study. On the NLET, the difference between the means for the kindergarten and first-grade performance was not significant; that is, no marked differences in performance for these students were detected by the NLET when they were grouped by class membership. For the ITBS, because two levels of the ITBS language section were used, statistical comparison of the group means could not be performed. However, content equivalence is assumed based on the skills objectives description in the teacher's guide, and the nearly equal standard deviations indicate the variability within the two groups as measured by the ITBS was comparable.

Because the goal of the project was to begin to identify a continuum of difficulty for language IRUS could handle, the students' raw scores for the ITBS

TABLE 3.4
Descriptive Statistics for Kindergartners and First-Graders on the ITBS Language Section and the NLET

	ITBS Language Section (29 Points)	NLET (39 Points)
Kindergartners		
M	21.25	32.18
SD	3.43	3.04
Range	7-27	25-37
(N = 60)		
First-Graders		
M	22.71	32.85
SD	3.88	2.64
Range	10-29	22-39
(N = 66)		
Kindergartners + First-Graders		
M	–	32.53
SD	–	2.84
Range	–	22-39
(N = 126)		

language section were converted to grade-equivalent scores using the tables provided in the ITBS teacher's guide. Grade-equivalent scores designate the year in school, preschool (P) through fourth grade (4) for these examinees, and the month in that academic year for a particular raw score. For example, P.9 indicates the ninth month of preschool; 4.3 indicates the third month of fourth grade, and so on. The grade-equivalent scores allowed for grouping the students by language proficiency rather than by age. Table 3.5 provides the descriptive statistics for the grade-equivalent bands.

When the students were grouped according to their grade-equivalent scores on the ITBS, a clear trend in their NLET performance appeared. There was steady improvement in NLET performance from the lowest grade-equivalent band to the highest, although there was considerable overlap in the range of NLET scores across grade-equivalent bands. (Unequal group sizes and variances prevented the use of analysis of variance, or ANOVA.)

Additionally, examination of the types of items missed by members of grade-equivalent bands did not reveal an aberrant pattern. That is, the members of the higher level bands did not miss different items from the members of the lower level bands; they simply missed fewer of the more difficult items. The findings suggest that the NLET captures developmental linguistic skills in transition and that establishing a continuum of difficulty for item types or items is possible. The continuum could eventually be expanded to provide the basis for a method of evaluating the natural language used by AI systems such as IRUS. To explore this possibility, an item analysis was conducted to identify potential response patterns and to attempt to isolate the linguistic sources of difficulty in the item types and items (see Baker et al., 1990, for details).

Establishing a Continuum of Linguistic Difficulty

IRUS, a sophisticated AI system that can process natural language, has demonstrated an ability to understand complex linguistic structures by responding to queries appropriately and consistently. The results of the testing effort reported

TABLE 3.5
NLET Descriptive Statistics for ITBS Grade-Equivalent Bands

ITBS Grade	N	NLET Mean	NLET SD	NLET Range
3.0 - 4.9[a]	17	34.47	2.27	31 - 38
2.0 - 2.9	20	32.85	1.57	29 - 35
1.0 - 1.9	37	32.59	2.97	22 - 37
K.O - K.9	42	32.00	2.95	26 - 37
P.O - P.9	10	30.60	3.20	26 - 35

[a]This band includes two students who had grade-equivalent scores of 4.9. The next highest grade-equivalent was 3.5. The 4.9 students were included in the third-grade band to avoid creating an artificial fourth-grade band.

TABLE 3.6
Difficulty Continuum in Order of Decreasing Difficulty

Group manipulation (recognition of the complete noun phrase referent for a pronoun).
Processing relative clauses joined by "or."
Processing prepositional phrase + relative clause.

---

here show that children of the ages tested also demonstrate a high degree of linguistic sophistication in terms of syntax and morphology. With the exception of items that involved group manipulation, items with a prepositional phrase and a relative clause, and those with two relative clauses joined by "or," none of the items appeared difficult on the basis of linguistic structure alone. Indeed, with the subjects tested there was a ceiling effect for most of the items. That is, the items measured linguistic structures that the majority of the students understood; thus, a continuum of difficulty cannot be developed for these items and the structures they include. However, we could begin to specify a continuum of difficulty based on NLET items that require group manipulation or the ability to process a prepositional phrase and a relative clause or relative clauses joined by "or."

Item analysis provided the basis for generating the continuum of difficulty reported in Table 3.6 in order of decreasing difficulty.

SUMMARY

Analysis of student performance on the NLET indicated that the test seems to tap some language skills in development within the population tested. An examination of the sources of difficulty in the item types and items yielded a tentative continuum of linguistic difficulty based on student performance on the NLET. NLET test specifications were developed from linguistic and functional analyses of the IRUS queries, so the beginning of a continuum of difficulty for the natural language understood by IRUS can be proposed. This comparison could be the first step in the development of a methodology for benchmarking AI systems to human performance.

In the pilot study, a natural language query system was referenced to the performance of kindergartners and first-graders. A criterion measure of language ability was used to allow for the grouping of students by national grade-equivalent norms. This grouping provided a means for benchmarking the NLET query system. The study yielded the beginning of a continuum of difficulty for the NLET understood by the query system. For the main study, a similar approach was taken. A criterion language measure was used so that national grade-level norms for reading could be established and would thus provide the necessary anchor for benchmarking.

The goal of this chapter has been to suggest a procedure for referencing the performance of intelligent computer systems, specifically a query system, to the performance of humans on the same task. The guiding question has been "Can we compare machine output to human performance?" Or perhaps more accurately, "Can we compare machine output of a specific kind to human performance of the same kind, and if so, how?" Our underlying assumption has been that performance-based measures can be used to reference query systems.

In general, the feasibility of using human benchmarking methodology to evaluate a specific natural language query system has been established. The "intelligence" of this application (IRUS) is low, as indicated by the equivalent performance of the query system compared to that of both kindergartners and first-grade students.

Caveats

Our approach has involved a benchmarking methodology that allows us to evaluate the natural language processing abilities of intelligent computer systems. The scale to which we have benchmarked the system's performance is a scale of human performance on language. We have outlined methods for determining overall system benchmarks and also methods for investigating differential system abilities, that is, benchmark levels for questions of different types. Due to the relatively small and clustered subject sample we had access to, we were not able to benchmark on a continuous scale nor were we able to determine the statistical significance of many of our results. Nevertheless, the general descriptive results presented here indicate that a human benchmark methodology can distinguish certain kinds of natural language processing abilities of intelligent computer systems. Given a larger and more diverse subject sample, it would be possible to rigorously substantiate benchmark values.

To place our efforts in context, it is important to realize that the work was exploratory and was conceived as part of a larger plan to develop a methodology for assessing a range of intelligent computer systems. Research with an expert system has also involved the use of a benchmarking methodology as an evaluation procedure (O'Neil et al., chap. 1). This growing body of work, which focuses on a human benchmarking approach to the assessment of intelligent computer systems, seems to suggest that we can compare system performance to human performance in a meaningful way using performance-based measures. Clearly the approach needs fine-tuning, but this study, as well as the others mentioned earlier, provides direction for researchers who are interested in a methodology for assessing intelligent computer systems.

ACKNOWLEDGMENTS

Thanks to Frances Butler, Jean Turner, Elaine Lindheim, Melinda G. Erickson, Tine Falk, Howard Herl, Younghee Jang, and Patricia Mutch for their assistance in various aspects of this research. Special appreciation to Dr. Susan Chipman of the Office of Naval Research for her support.

This research was supported by contract number N00014-86-K-0395 from the Defense Advanced Research Projects Agency (DARPA), administered by the Office of Naval Research (ONR), to the UCLA Center for the Study of Evaluation/Center for Technology Assessment. However, the opinions expressed do not necessarily reflect the positions of DARPA or ONR, and no official endorsement by either organization should be inferred. This research also was supported in part under the Educational Research and Development Center Program cooperative agreement R117G10027 and CFDA catalog number 84.117G as administered by the Office of Educational Research and Improvement, U.S. Department of Education. The findings and opinions expressed in this report do not reflect the position or policies of the Office of Educational Research and Improvement or the U.S. Department of Education.

REFERENCES

Baker, E. L., Turner, J. L., & Butler, F. A. (1990). *An initial inquiry into the use of human performance to evaluate artificial intelligence systems.* Los Angeles: University of California, Center for Technology Assessment/Center for the Study of Evaluation.

Bates, M., Stallard, D., & Moser, M. (1986). The IRUS transportable natural language database interface. In L. Kerschberg (Ed.), *Expert database systems* (pp. 617–630). Menlo Park, CA: Cummings Publishing.

Defense Advanced Research Projects Agency, Information Science and Technology Office (DARPA). (1989, February). *Proceedings, speech and natural language workshop.* San Mateo, CA: Morgan Kaufman Publishers.

Hieronymus, A. N., & Hoover, H. D. (1986). *Manual for school administrators, levels 5 & 6, forms G/H, Iowa Tests of Basic Skills (ITBS).* Chicago: Riverside Publishing Company.

Hieronymus, A. N., Hoover, H. D., & Lindquist, E. F. (1986). *Teachers guide, early primary battery, levels 5 & 6, Iowa Tests of Basic Skills (ITBS).* Chicago: Riverside Publishing Company.

Jacobs, P. S. (1990). Two hurdles for natural language systems. In R. Freedle (Ed.), *Artificial intelligence and the figure of testing* (pp. 257–277). Hillsdale, NJ: Lawrence Erlbaum Associates.

Lehnert, W., & Sundheim, B. (1991). An evaluation of text analysis technologies. *AI Magazine, 12*(3), 81–94.

4 Assessment of Explanation Systems

Johanna D. Moore
University of Pittsburgh

The need for explanation capabilities in intelligent systems, and especially expert systems, has been widely recognized. Most commercially available expert system building tools provide at least rudimentary explanation capabilities. Yet, few rigorous and thorough evaluations have been attempted. This is in part due to the fact that an explanation system is not stand-alone, but occurs as a component of a larger, intelligent system and the quality of a system's explanations depends on factors outside the control of the explanation component. Criteria and methods for evaluating explanation systems are only now beginning to emerge.

This chapter describes the goals of explanation systems and discusses three methods for determining if those goals are being achieved: assessing the impact of the system on the user's satisfaction or performance, comparing human–machine explanatory interactions to human–human advisory interactions, and assessing a system according to a set of evaluation criteria. The first two methods have been employed and the results of case studies using these methods are discussed. The third method relies on developing appropriate evaluation criteria and a set of recently proposed criteria is presented. These criteria include metrics such as the cost of providing a system with explanation capabilities and the benefits gained from making this investment. Current research indicates that providing a system with explanation capabilities requires substantially more effort during system development, but these costs are more than recovered during the software life cycle because the kinds of architectures that support explanation also facilitate knowledge acquisition and system maintenance. The chapter concludes with a brief look at new developments in explainable system architectures.

EXPLANATION SYSTEMS: DEFINITION AND GOALS

As the name suggests, an explanation system presents its users with explanatory information in response to user requests. This system is not stand-alone, but rather a component of a knowledge-based system such as an expert system (a computer program that employs a representation of knowledge and a set of inference procedures to solve problems that normally require human expertise) or an intelligent tutoring system (a system that teaches the subject matter in a domain using a representation of the subject matter, and pedagogical strategies for reducing the difference between student and expert performance in the domain). The explanation component of such a system is intended to elucidate the intelligent system's domain knowledge and behavior by producing explanations such as definitions of terms, justification of results, and comparisons of alternate methods for solving problems or achieving goals.

The critical need for explanation has been voiced by expert system builders and the intended user community alike. In a study by Teach and Shortliffe (1984) in which physicians were asked to rank 15 capabilities of computer-based consultation systems in order of importance, the ability "to explain their diagnostic and treatment decisions to physician users" was rated as the most essential of the capabilities surveyed. Third on the list was the ability to "display an understanding of their own medical knowledge." The importance of explanation capabilities to these users is underscored by the fact that the capability to "never make an incorrect diagnosis" was ranked 14 out of 15 capabilities surveyed! This suggests that explanation capabilities are not only desirable, but crucial to the success of knowledge-based systems.

Researchers have designed explanation systems with a variety of goals in mind. For end users, these goals include the following:

- **Information:** Explanation systems offer access to the considerable knowledge available in an intelligent system.
- **Education:** Explanation can be used to educate users about the problem domain.
- **Acceptance:** The advice of intelligent systems will not be accepted unless users believe they can depend on the accuracy of that advice. Systems that can display their domain knowledge and justify the methods used in reaching conclusions are considered more acceptable to the user community (Teach & Shortliffe, 1984).
- **Assess appropriateness of system for task:** The scope of intelligent systems is typically narrow. Swartout (1983) pointed out that an explanation facility can help a user discover when a system is being pushed beyond the bounds of its expertise by presenting the methods being employed and rationales for employing them in a given situation.

In addition to the benefits offered to end users of systems, an explanation facility can also be useful for system developers. The following is an additional goal of an explanation system:

- **Aid in debugging and maintaining system:** Because a textual or graphical explanation of a system's knowledge or behavior offers a viewpoint different from the code that implements it, system developers find that such explanations can aid in locating bugs and also offer a form of documentation that is helpful in maintenance.

DIFFICULTIES IN EVALUATING EXPLANATION SYSTEMS

Research in the area of expert system explanation has led to the understanding that explanation capabilities cannot be added to a system as an afterthought. The capability to support explanation imposes stringent requirements on the design of an expert system, and it can be difficult, if not impossible, to endow a system with the capability to produce adequate explanations unless those requirements are taken into consideration when designing the system (Clancey, 1983b; Swartout, 1983). Retrofitting an existing intelligent system with explanation capabilities inevitably means redesigning it.

The range of user questions an explanation facility is able to handle and the sophistication of the responses it can produce depend on two knowledge sources: (a) knowledge about the domain and how to solve problems in that domain as represented in the intelligent system's knowledge base, for example, definitions of terminology, justifications for system actions; and (b) knowledge about how to construct an adequate response to a user's query in some communication medium, for instance, natural language and/or graphics. The latter includes methods for interpreting the user's questions and strategies for choosing information relevant to include in answers to different types of questions.

In the expert systems area, researchers have determined that many of the inadequacies noted in the explanations stem from limitations in the systems' knowledge bases. Until recently, expert system architects have concentrated their efforts on the problem-solving needs of systems. They designed knowledge bases and control mechanisms best suited to performing an expert task. Because of this, the explanation capabilities of these systems were limited to producing procedural descriptions of *how* a given problem is solved. Knowledge needed to justify the system's actions, explain general problem-solving strategies, or define the terminology used by the system was simply not represented and therefore could not be included in explanations (Clancey, 1983b; Swartout, 1983). Deficiencies in the knowledge representation lead to other types of inadequacies, rendering systems inflexible, insensitive, and unresponsive to users' needs. To provide different explanations to different users in different situations or in response to requests for elaboration, the system must have a rich representation of its knowledge, including abstract strategic knowledge as well as detailed knowledge, a rich terminological base, and causal knowledge of the domain (Swartout & Moore, 1993).

The fact that the explanation capabilities of a system are tightly coupled to the knowledge base and reasoning component of that system leads to a fundamental problem for those interested in the assessment of explanation systems. An explanation facility cannot be evaluated on its own. It must be evaluated in the context of the larger intelligent system of which it is a part. It would be unreasonable to fault the explanation facility for its inability to respond to users' requests for rule justifications, when the knowledge need to justify rules is not present in the system's knowledge base. Thus, evaluating an explanation facility is a complex task. It inevitably involves assessing the entire intelligent system, dissecting the system into components so that blame may fairly be assigned to the offending component(s), and defining the types of knowledge needed to produce explanations, all in order to gain an understanding of what limitations could stem from the knowledge base and what limitations stem from the explanation component itself.

In the following, I describe three methods that could be used to evaluate explanation systems. The first two methods have been employed, and I describe two specific cases in detail. These case studies had the goal of identifying specific problems with existing explanation systems and uncovering the limitations behind the perceived inadequacies. Such studies led researchers to an understanding of the requirements that the need to provide explanations places on the knowledge representation and reasoning processes of an intelligent system. This chapter describes the limitations identified in the case studies as well as the general principles that came from in-depth analyses of these limitations. Finally, it discusses current research efforts in explanation technologies that are attempting to provide applications builders with an explainable intelligent system framework by capturing the knowledge needed to support explanation and structuring that knowledge so that it is available to the explanation facility. Although such frameworks are still exploratory, and the feasibility of providing domain-independent reasoning and explanation facilities remains to be demonstrated, many promising results are emerging from this work. One intriguing possibility from the perspective of evaluation is that the requirements posed by an explainable intelligent system framework may in fact suggest a way to evaluate whether a system can readily accept an explanation module.

METHODS FOR EVALUATION

Given the nature of explanation systems and their tight coupling to other components of intelligent systems, I believe there are three possible methods for assessing explanation capabilities. Note that these three methods correspond to three

different definitions of "technology assessment." The first method defines technology assessment in terms of the system's impact on the user, either the user's
acceptance of the system or the system's impact on the user's performance. The
second method defines technology assessment as a measure of similarity between
the system's behavior and the human behavior the system is attempting to mimic.
The third method defines technology assessment as a rating based on a set of
criteria aimed at evaluating the mechanism by which explanations are produced,
the explanations themselves, and the cost of providing explanations.

Assess Impact on User

The purpose of an explanation component is to facilitate the user's access to the
information and knowledge stored in the intelligent system. Thus one way to
assess an explanation component is to assess the impact of the explanation
component on the user's behavior or satisfaction with the system. This can be
done using direct methods, such as interviewing users to determine what aspects
of the system they find useful and where they find inadequacies, or by indirect
methods, which measure users' performance after using the system or monitor
usage of various facilities.

Interviewing Users.   One of the best sources of assessment information is the
user population.[1] If users feel that an explanation facility is meeting their needs,
then the explanation facility is successful by one important measure, namely user
acceptance. Experience indicates that users are more likely to report frustrations
and inadequacies with the system, but this is also useful information. As is
discussed in depth in the section on MYCIN, determining what architectural
features of the system are responsible for the inadequacies identified by users
points up areas where systems must be improved and, in expert system explanation, has inspired research efforts to alleviate the limitations (e.g., Clancey &
Letsinger, 1984; Moore & Swartout, 1989).

Monitoring Usage of Explanation Facility.   Another telling assessment of
any automated tool is whether or not users actually avail themselves of the tool
and whether or not their usage is successful; that is, they are able to get the
information they seek or are able to make the system perform the task they
desire. To my knowledge, a study of this kind has not been done specifically in
the area of explanation systems, but such studies have been done in the area of
help systems. For example, an empirical study of usage of the Symbolics Docu-
MENT EXAMINER, an on-line documentation system that supports keyword
searchers, indicated that a substantial number of interactions with the system
ended in failure, especially when users were inexperienced (Young, 1987).

---

[1]Alessi and Trollip (1991) provided several interesting structural suggestions for interviewing.

Impact on User's Performance on Task.   Another way to assess the contribution of an explanation component is to determine how the explanation capabilities of the system contribute to users' effectiveness in using the system to achieve their goals. For example, if the system is intended to instruct users about how to perform some task, then it should be possible to design a simple experiment that assesses the explanation component. One way to set up such an experiment would be to have two groups of users. One group would use a version of the system with full explanation capabilities. The other group would be given the system without explanation capabilities. Both groups' performance on the task would be measured before and after using the system. If the explanation capabilities of the system are effective, we would expect the group that used the system with explanation capabilities to show greater improvement in task performance, greater retention of skills, or greater transfer of knowledge to related tasks.

Comparison to Human–Human Advisory Interactions

Fischer (1987) claimed that "human assistance, if available on a personal level, is in almost all cases the most useful source of advice and help" (p. 179). One reason he cited for this is that most information and advice-giving systems require that users know what they are looking for when they approach such a system. However, studies of naturally occurring advisory interactions indicate that advice seekers often require assistance in formulating a query (Finin, Joshi, & Webber, 1986; Pollack, Hirschberg, & Webber, 1982).

Thus, another way to assess explanation facilities is to compare them to the ideal "explanation system," that is, a human explainer, and to determine what capabilities of human–human explanatory interactions are present or absent from the system being evaluated. In my own work on expert system explanation, I have performed such a comparison, which is discussed in detail later.

Evaluate System Against Set of Criteria

A third approach for evaluating explanation systems would involve devising a set of evaluation criteria and rating systems accordingly. This requires not only devising the set of evaluation metrics, but also devising techniques for measuring the system along each dimension in an objective fashion. In the section on Method 3, I discuss one set of evaluation criteria that has recently been proposed. That section points out that devising routines to measure how a system fares against the criteria may not be an easy task, and for certain of the criteria, the measurement may inevitably remain subjective.

The next two sections describe case studies using the first two methods. A set of evaluation criteria proposed for use with the third method is then discussed.

METHOD 1: EVALUATING MYCIN'S EXPLANATION FACILITY

One of the first studies that attempted to evaluate an explanation facility was done in the context of MYCIN, a rule-based medical consultation system designed to provide advice regarding diagnosis and therapy for infectious diseases (Shortliffe, 1976; Buchanan & Shortliffe, 1984). The rules encoding MYCIN's medical knowledge are composed from a small set of primitive functions that make up the rule language. Associated with each of the primitive functions is a template to be used when generating explanations. The English translation of a sample MYCIN rule is shown in Fig. 4.1.

In MYCIN, a consultation is run by backward chaining through applicable rules, asking questions of the user when necessary. As a consultation progresses, MYCIN builds a *history tree* reflecting the goal/subgoal structure of the executing program. Explanations are produced from the history tree using the templates to translate the sequence of rules that were applied to reach a conclusion.

To study MYCIN's explanation facility, several scenarios in which MYCIN produced inadequate responses to questions asked by users were analyzed in order to determine the reasons for the suboptimal performance (Buchanan & Shortliffe, 1984). As a result of this study, the implementors of MYCIN discovered several problems with the explanation facility and were able to identify limitations in the system's architecture that were responsible for these problems.

The following problems were identified by the implementors:

• MYCIN could not answer some types of questions that users of the system wished to ask. Most notably, MYCIN could not produce justifications of the rules it used in making a diagnosis; that is, MYCIN could not answer questions of the form "Why does the conclusion of a rule follow from its premises?"

• MYCIN could not deal with the context in which a question was asked. MYCIN had no sense of dialogue, so each question required full specification of the points of interest without reference to earlier exchanges.

IF: 1. the infection that requires therapy is meningitis,
    2. only circumstantial evidence is available for this case,
    3. the type of meningitis is bacterial,
    4. the age of the patient is greater than 17 years old, and
    5. the patient is an alcoholic,

THEN: there is evidence that the organisms that might be causing the infection
       are *diplococcus-pneumoniae* (.3) or *e. coli* (.2)

FIG. 4.1. A MYCIN rule with implicit knowledge.

• MYCIN often misinterpreted the user's intent in asking a question.The study identified examples of simple questions with four or five possible meanings depending on what the user knows, the information currently available about the patient under consideration, or the content of the earlier discussions.

The first problem reflects the lack of sufficient knowledge to support explanation, whereas the remaining two stem from the limitations of ad hoc explanation techniques that fail to handle the linguistic complexities of explanation generation.

Impoverished Knowledge Bases

In attempting to adapt MYCIN for use as a tutoring system, Clancey examined MYCIN's rule base and found that individual rules served different purposes, had different justifications, and were constructed using different rationales for the ordering of clauses in their premises (Clancey, 1983b). However, these purposes, justifications, and rationales were not explicitly included in the rules; therefore many types of explanations simply were not possible, and thus, user questions that required such explanations as responses could not be addressed. In particular, there were three important types of explanation that MYCIN, and other early systems that generated explanations by translating their code, could not produce.

Justifications.   Early systems could not give justifications for their actions. These systems could produce only procedural descriptions of what they did. They could not tell the user why they did it or why they did things in the order that they did them.

For example, consider the rule shown in Fig. 4.1 and suppose that the user wishes to know why the five clauses in its premise suggest that the organism causing the infection may be *diplococcus* or *E. coli,* MYCIN cannot explain this because the system knows no more about the association between the premises and conclusion than what is stated in this rule. In particular, MYCIN does not know that Clauses 1, 3, and 5 together embody the causal knowledge that if an alcoholic has bacterial meningitis, it is likely to be caused by *diplococcus.* Furthermore, MYCIN does not understand that Clause 4 is a *screening clause* that prevents the system from asking whether the patient is an alcoholic when the patient is not an adult—thus making it appear that MYCIN understands this social "fact." However, MYCIN does not explicitly represent, and therefore cannot explain, this relationship between Clauses 4 and 5. Even worse, not knowing that the system makes this assumption may lead the user to infer that age has something to do with the type of organism causing the infection.

Furthermore, the order in which rules are tried to satisfy a particular goal will affect the order in which subgoals are pursued. When a goal, such as determining the identity of the organism, is being pursued, MYCIN invokes all of the rules concerning the identity of the organism in the order in which they appear in the rule base. This order is determined by the order in which the rules were entered into the system! Thus MYCIN cannot explain why it considers one hypothesis before another in pursuing a goal. In addition, because attempting to satisfy the premises of a rule frequently causes questions to be asked of the user, MYCIN cannot explain why it asks questions in the order it does because this too depends on the ordering of premises in a rule and the ordering of rules in the knowledge base. As Clancey (1983, p. 220) pointed out "focusing on a hypothesis and choosing a question to confirm a hypothesis are not arbitrary in human reasoning" and thus users will expect the system to be able to explain why it pursues one hypothesis before another and will expect questions to follow some explicable line of reasoning.

Explications of General Problem-Solving Strategy.   The knowledge needed to explain general problem-solving strategies was not explicitly represented in early systems. In MYCIN, because several different types of knowledge are confounded in the uniformity of the rule representation, it is difficult to identify and explain MYCIN's overall diagnostic strategy.

Again, consider the rule shown in Fig. 4.1. Another hidden relationship exists between Clauses 1 and 3. Clearly, bacterial meningitis is a type of meningitis, so why include Clause 1? The ordering of Clauses 1 and 3 implicitly encodes strategic knowledge of the consultation process. The justification for the order in which goals are pursued is implicit in the ordering of the premises in a rule. The choice of ordering for the premises is left to the discretion of the rule author and there is no mechanism for recording the rationale for these choices. This makes it impossible for MYCIN to explain its general problem-solving strategy; that is, that it establishes the infection as meningitis before it determines if it is bacterial meningitis because it is following a refinement strategy of diagnosing the disease (Hasling, Clancey, & Rennels, 1984).

Definitions of Terminology.   Terminological knowledge was also not explicitly represented in MYCIN or other early systems. Users who are novices in the task domain will need to ask questions about terminology to understand the system's responses and to be able to respond appropriately when answering the system's questions. Furthermore, experts may want to ask such questions to determine whether the system's use of a term is the same as their own. Because the knowledge of what a term means is implicit in the way it is used, the system is not capable of explaining the meaning of a term in a way that is acceptable to users. An effort to explain terms by examining the rule base of an expert system (Rubinoff, 1985) has been only partially successful because so many different types of knowledge are encoded into the single, uniform rule formalism. This makes it difficult to distinguish definitional knowledge from other types of knowledge.

Summary. As we have seen, rules and rule clauses incorporate many different types of knowledge, but the uniformity of the rule representation obscures their various functions, thus making comprehensible explanation impossible. Much of the information that goes into writing the rules or program code that makes up an expert system—including justification, strategic information, and knowledge of terminology—is either lost completely or made implicit to the extent that it is no longer available to be explained. Both Swartout (1983) and Clancey (1983b) argued that the different types of knowledge (definitional, world facts, causal, strategic) must be separately and explicitly represented if systems are to be able to explain their strategy and justify their conclusions. Consequently, later efforts have addressed the problem of capturing the knowledge and decisions that went into writing the program and explicitly representing this information so that it will be available for explanation. This research is briefly discussed here. For a more detailed discussion of knowledge representation issues, and a description of the explainable expert systems approach, see Swartout and Moore (1993) and Swartout, Paris, and Moore (1991).

Inadequate Natural Language Techniques

The remaining two problems identified by the MYCIN implementors in their analysis of inadequate responses stem from limitations in the natural language capabilities of MYCIN. Specifically, question understanding and interpretation procedures are limited, thus restricting the kinds of questions that may be asked and the manner in which they must be phrased. To avoid the difficult problems of inferring the user's intent in asking a question, MYCIN interprets a user's "Why?" query in only one way, even though it could have a variety of meanings, such as "Why is it important to determine _____ ?", "Why did you ask about that instead of _____ ?", "Why do you ask that now?", or "Why does the conclusion follow from the premises?" All of these interpretations are valid questions about the system's knowledge and behavior, yet MYCIN always assumes the first interpretation and does not allow the other questions to be asked.

For example, consider the explanation produced by MYCIN in Fig. 4.2. When the user asks "Why?" the second time, the system assumes that the user is asking "Why is it important to determine the category of ORGANISM-1?" But the user may really be asking a very different type of question, namely, "Why is it the case that a gram negative, facultative rod acquired in a hospital setting is likely to be pseudomonas?"[2] We have already seen that the causal knowledge needed to answer this question is not represented in MYCIN. But even if it were, MYCIN could not determine what why-question was being asked in a given context because MYCIN does not view the explanation session as an ongoing dialogue. Each question–answer pair is viewed independently, and references to

---

[2]This example is adapted from an example found in Davis and Lenat (1982).

Was the infection with ORGANISM-1 acquired while the patient was hospitalized?

**WHY?

[i.e., WHY is it important to determine whether or not the infection with ORGANISM-1 was acquired while the patient was hospitalized?]

[1.0] This will help to determine the identity or ORGANISM-1.

It has already been established that

[1.1] the gram stain or ORGANISM-1 is gramneg, and
[1.2] the morphology of ORGANISM-1 is rod, and
[1.3] the aerobicity of ORGANISM-1 is facultative

Therefore, if

[1.4] the infection with ORGANISM-1 was acquired while the patient was hospitalized

Then

there is a weakly suggestive evidence (.2) that the identity of the organism is pseudomonas

[RULE050].

**WHY?

[i.e., WHY is it important to determine the identity of ORGANISM-1?]

FIG. 4.2.   A sample MYCIN explanation.

previous portions of the dialogue can be made only in stilted and artificial ways. A model of explanation that addresses this problem is described here.

Summary.   The experience with the MYCIN study shows that asking users to indicate unsatisfactory explanation behavior is a very useful method for evaluation. This study pointed out several scenarios where MYCIN's explanation facility did not behave as users expected or could not give an explanation they desired. Analyzing these scenarios led MYCIN's implementors and other researchers to perform in-depth studies to uncover the architectural limitations that were responsible for MYCIN's inadequate explanations. These studies are examples of what Littman and Soloway (1988) called *internal evaluation,* a technique in which assessors perform a detailed analysis of a system's architecture in order to determine the relationship between the system's architectural features and its behavior. Internal evaluations of MYCIN and other first generation expert systems have led to a deeper understanding of the architectural requirements that explanation places on a knowledge-based system, and have spurred research aimed at improving the problems. See Swartout and Moore (1993) for a critical survey of expert system architectures and their ability to support explanation.

METHOD 2: COMPARISON TO HUMAN–HUMAN INTERACTIONS

My own work in expert system explanation was motivated by an observation of a great disparity between what analyses of naturally occurring advisory dialogues reveal, on the one hand, and the explanation facilities that current systems provide and the assumptions they make about how users interact with experts, on the other. Most systems make the tacit assumption that the explanations they produce will be understood by their users. However, in human–human advisory situations, people almost always ask follow-up questions! Expert and advice-giving systems are expected to provide solutions and advice to users faced with real problems in complex domains. They often produce complex multisentential responses, such as definitions of terms, justification of results, and comparisons of alternate methods for solving problems. Users must be able to ask follow-up questions if they do not understand an explanation or want further elaboration. Answers to such questions must take into account the dialogue context.

What the Data Reveal

In a study of a "naturally occurring" expert system, Pollack et al. (1982) found that user–expert dialogues are best viewed as a negotiation process in which the user and expert negotiate the statement of the problem to be solved in addition to a solution that the user can understand and accept. In my own work on explanation, I examined samples of naturally occurring dialogues from several different sources: tape-recordings of office-hour interactions between first-year computer science students and teaching assistants, protocols of programers interacting with a mock program enhancement advisor, and transcripts of electronic dialogues between system users and operators taken from Robinson (1984). The following is a portion of a dialogue extracted from the transcripts collected:

TEACHER: OK, so what is it, it's using stacks, right?
STUDENT: Yea, well, cause, um, aren't we supposed to use linked lists?
TEACHER: You don't have to use linked lists. You don't.
STUDENT: But OK. You said stacks, right?
TEACHER: In LISP we implemented stacks as linked lists. In C we can implement stacks as an array.
STUDENT: Wait, in LISP . . .
TEACHER: In LISP, we implemented, past tense, implemented stacks as linked lists. Right? In C, we can do it anyway we want. We can implement it as linked lists or as arrays, uh, I don't know any obvious data structures after that. But, um, you can use a linked list or an array. I would use an array, personally.

In this dialogue, the student does not understand the difference between the general concept of a stack as a way of managing data and its implementation

110

using a particular data structure. The teacher thinks she has cleared up the student's misunderstanding with her first explanation, but in fact she has not, as indicated by the student's "Wait" and hesitation. The teacher enhances her earlier response by emphasizing the point she made in the previous explanation, and then elaborating on the notion that a stack can be implemented using various data structures. Similarly, automated systems must be able to offer further elaborations of their responses or alternate explanations, even when the user is not very explicit about which aspect of the explanation was not clear.

An analysis of dialogues (Moore, 1989) such as this one led to the following observations:

- *Users frequently do not fully understand the expert's response.* Users rarely stated a problem, received a result or explanation, and then left satisfied that they understood and accepted the expert's explanation. The expert frequently found the need to define terms or establish background information in response to feedback that the listener did not completely understand the response.

- *Users frequently ask follow-up questions.* Users frequently requested clarification, elaboration, or re-explanation of the expert's response.

- *Users often don't know what they don't understand.* Users frequently could not formulate a clear follow-up question. In many cases, the follow-up question was vaguely articulated in the form of mumbling, hesitation, repeating the last few words of the expert's response, or simply stating "I don't understand." Often the expert did not have much to go on, but still was required to provide further information.

- *Experts do not have a detailed model of the user.* From the dialogues I examined, it is clear that experts do not have a complete and correct model of their users. Although we can safely assume that experts have some model of the users, it seems that because many and varied users are likely to seek their help, this model is more likely to be a stereotypic model that may be incomplete or incorrect for any given hearer than a detailed model of any individual. Yet, as the dialogues showed, the user and the expert are able to communicate effectively.

The State of Conventional Explanation Systems

Studies such as these and many others (Cawsey, 1993; Finin et al., 1986; Suchman, 1987) show that explanation requires dialogue, yet few intelligent systems participate in a dialogue with their users. The explanation facilities of most current systems can be characterized in the following ways:

- **Unnatural:** Explanation generation does not employ linguistic knowledge about how texts should be constructed. The unnatural structure of the resulting texts often obscures the important points in an explanation.
- **Inextensible:** New explanation strategies cannot be added easily.

- **Unresponsive:** The system cannot answer follow-up questions or offer an alternative explanation if a user does not understand a given explanation.
- **Insensitive:** Explanations do not take the context into account. Each question and answer pair is treated independently.
- **Inflexible:** Explanations can be presented in only one way.

Generally, these problems stem from the fact that explanation generation has not been considered as a problem requiring its own expertise and worthy of an architecture that supports a sophisticated problem-solving activity. More specifically, these problems stem from limitations in several areas as outlined here.

Many expert systems produce explanations from a trace of the expert system's line of reasoning. Much effort has gone into identifying clever strategies for annotating, pruning, traversing, and translating the execution trace to produce "good" explanations (e.g., McKeown, 1988; Wallis & Shortliffe, 1984; Weiner, 1980). However, there are several problems with this approach. First, it places much of the burden of producing explanations on expert system builders and their ability to structure the rules or program code in a way that will be understandable to users who are knowledgeable about the task domain. For example, in MYCIN, an attempt was made to make each rule an independent "chunk" of medical knowledge reflecting a complete, coherent explanation. In addition, as other researchers (Clancey, 1983b; Davis & Lenat, 1982; Pavlin & Corkill, 1984; Swartout 1983; Webber & Joshi, 1982) have noted, the computationally efficient reasoning strategies used by expert systems to produce a result often do not form a good basis for understandable explanations. Experience has shown that explanations produced by paraphrasing the system's execution trace correspond too literally to the program structure, which is dictated, at least in part, by implementation considerations that may obscure the underlying domain-related reasoning. Moreover, there is no reason to assume that a simple paraphrase of the program's execution will produce an explanation that conforms to the linguistic conventions dictating discourse structure and coherence.

Consider the explanation in Fig. 4.3, which was produced by NEOMYCIN. As is discussed further, NEOMYCIN (Clancey & Letsinger, 1984) was developed as part of an effort to build an intelligent tutoring system to instruct medical students in diagnosis. For this purpose, it was necessary for the system to be able explain its overall diagnostic strategy. Recall that MYCIN could not answer questions of this type because of limitations in its knowledge base. In NEO-MYCIN domain-independent, problem-solving knowledge is represented explicitly and separately from knowledge about the domain. Because its explanations are based on a more explicit representation, NEOMYCIN is able to produce explanations of its problem-solving strategies that are a significant improvement over those of MYCIN. For example, NEOMYCIN is able to give abstract explanations of its general problem-solving strategy (see Hasling et al., 1984, for details). However, the explanation in Fig. 4.3 demonstrates that, even with the

---

"HOW 19.0?

[i.e., How are we trying to decide whether Mary has tuberculosis?]

We did the following:

[20.1] Applied a rule mentioning a causal precursor to
      tuberculosis (tb risk factor, rule 525), and
[20.2] Gave thorough consideration to any recent findings.
[20.3] Applied a rule mentioning general evidence for
      tuberculosis (focal neurological signs; rule 366), and
[20.4] Gave thorough consideration to any recent findings.
[20.5] Applied a rule mentioning general evidence for
      tuberculosis (compromised host; rule 309), and
[20.6] Gave thorough consideration to any recent findings.

We are now:

[20.7] Applying a rule mentioning general evidence for tuberculosis
      (a history of granulomas on biopsy of the liver, lymph node, or
      other organ; rule 002).

We will next:

[20.8] Give thorough consideration to any recent findings.

[back to question 14 ....]

FIG. 4.3.   A sample NEOMYCIN explanation.

improved knowledge base of NEOMYCIN, simply paraphrasing the system's reasoning often produces text that is unsuitable for users. The problem with this explanation is that its structure corresponds too literally to the structure of the method for achieving the task, that is, applying one rule after another. Whereas the explanation repeats that it "gave thorough consideration to any recent findings" four times, the explanation does not make clear the overall strategy of applying rules that strongly contribute to the current hypothesis (e.g., causal precursors) before applying rules that are weaker indicators (e.g., general evidence). What is needed is a more sophisticated explanation strategy that could recognize the similarity in the four rule applications and structure this information accordingly. This example is symptomatic of the more general problem that simply paraphrasing the system's knowledge base or reasoning trace is not guaranteed to produce understandable explanations, regardless of how well that knowledge base is structured. In order to produce natural language explanations that are coherent to human users, an explanation facility must have linguistic knowledge about discourse structure and strategies for employing that knowledge to achieve its explanatory goals.

Another problem is that the explanation components of conventional expert systems are difficult to extend. Because explanation is typically implemented in ad hoc procedures or large collections of rules, it is difficult to understand how adding new rules or procedures will interact with existing facilities. In general, answering a new type of question involves coding new procedures or building new templates from scratch. Little, if any, of the existing code is useful. Moreover, expert systems are becoming more complex as system builders augment their knowledge bases to include the underlying support knowledge needed to allow systems to answer a broader range of questions about their domain knowledge and behavior. The knowledge bases of newer expert systems separate different types of knowledge (Clancey & Letsinger, 1984; Swartout et al., 1991) and represent knowledge at various levels of abstraction (Patil, Szolovits, & Schwartz, 1984), and thus confront explanation generators with an array of choices about what information to include in an explanation and how to present the information that was unavailable in the simple knowledge bases of earlier systems. Meeting the challenge posed by these richer knowledge bases will require more sophisticated and linguistically motivated explanation generators.

Researchers in computational linguistics have addressed the issues of selecting information from a complex knowledge base and organizing it into a text that adheres to the conventions of discourse structure (cf. Appelt, 1985; McCoy, 1989; McKeown, 1985; Paris, 1991b; Reichman-Adar, 1984; Weiner, 1980). Results from this research have been successfully incorporated into recent explanation systems, for example (Cawsey, 1993; Moore & Swartout, 1989; Paris, 1991a; Suthers 1991; Wolz 1990), to enable them to produce natural explanations.

However, few systems are capable of responding to follow-up questions. The problem is that most intelligent systems that respond to user's questions view generating responses as a *one-shot process*. That is, they assume they will be able to produce an explanation that the user will find satisfactory in a single response. This one-shot approach is inconsistent with analyses of naturally occurring advisory dialogues. Moreover, if a system has only one opportunity to produce a text that achieves the speaker's goals without over- or under-informing, boring, or confusing the listener, then that system must have an enormous amount of detailed knowledge about the listener. Taking the one-shot approach has led to a view that improvements in explanation will follow from improvements in the *user model*. Recognizing this need, researchers have studied how to build user models, for example (Chin, 1989; Kass, 1991; Kobsa, 1989; Mastaglio, 1990; Rich, 1989; Wu, 1991), and how to exploit them to produce the "best" possible answer in one shot. Considerable effort has been expended on building complex user models, containing large amounts of detailed information about a user, including the user's goals and plans, attitudes, capabilities, preferences, level of expertise, beliefs, and beliefs about the system's beliefs (and other such mutual beliefs). From these models, a system then attempts to generate the "best possible" answer for that particular user. Already, systems employing such models have demonstrated that a user model can be used to guide a generation system in producing answers that appear to be appropriately tailored to the user (Appelt, 1985; Kass, 1991; McCoy, 1989; Paris, 1991b; van Beek, 1987; Wolz, 1990).

Whereas the quality of explanations can be demonstrably improved by employing a user model, a system that is critically dependent on such a model will not suffice. Sparck Jones (1989) questioned whether it is even feasible to build complete and correct user models. To date, no robust system for automatically acquiring these complex and detailed user models exists, and hand-crafting them is time consuming and error prone. The completeness and accuracy of a user model cannot be guaranteed. Thus, unless mechanisms are developed by which systems can dynamically acquire and update a user model by interacting with the user, the impracticality of building user models will prevent much of the work on tailoring from being successfully applied in real systems. Some researchers have begun to develop tools to aid systems in acquiring user models. For example, the General User Modeling Acquisition Component (GUMAC) (Kass, 1991) provides a set of domain-independent acquisition rules that allows a system to acquire a model of the user by drawing inferences from the user's utterances during a dialogue with an expert system. Although such tools show promise, systems that rely on correct and complete user models are likely to be brittle. Indeed, most of Kass' acquisition rules require that the user and system be able to carry on an initial dialogue (for example, to gather data or establish the problem to be solved) during which the user model is being acquired. Thus the system must be able to carry on a dialogue with the user in order to acquire the user model. Clearly, the system must be able to communicate without a complete and correct model if Kass' system is to be feasible.

More importantly, by focusing on user models, researchers have ignored the rich source of guidance that people use in producing explanations, namely feedback from the listener (Ringle & Bruce, 1981).

What Is Needed in a Dialogue System

In my own work, I have developed a model of explanation in which feedback from the user is an integral part of the explanation process. In designing this model, I identified the capabilities that a system must possess in order to participate in a dialogue.

Accept Feedback From the User.   To be responsive to a user's feedback, a system must allow the user to provide that feedback. Many systems do not have a means for allowing the user to indicate dissatisfaction with a given explanation. The user who cannot ask one of a prescribed set of follow-up questions in a prescribed form is without recourse.

The System Must Understand its Own Explanations.   In order to provide elaborating explanations, clarify misunderstood explanations and respond to follow-up questions in context, a system must view the explanations it produces as objects to be reasoned about later. In particular, detailed knowledge about how the explanation was "designed" must be recorded, including: the goal structure of the explanation, the roles individual clauses in the text play in responding to the user's query, how the clauses relate to one another rhetorically, and what assumptions about the listener's knowledge may have been made. Without this knowledge, a system cannot be sensitive to dialogue context or responsive to the user's needs.

Interpret Questions Taking Previous Explanations Into Account.   When participating in a dialogue, a system must realize that the same question can mean different things in different contexts; that is, the system must be able to interpret questions taking into account what the user knows, the information available about the current problem-solving situation, or the content of the previous discussion.

Have Multiple Strategies for Answering a Question of a Given Type.   Finally, making oneself understood often requires the ability to present the same information in multiple ways or to provide different information to illustrate the same point. Current systems are inflexible because they typically have only a single response strategy associated with each question type, instead of the sophisticated repertoire of discourse strategies that human explainers utilize. Without multiple strategies for responding to a question, a system cannot offer an alternative response even if it understands why a previous explanation was not satisfactory.

Based on these requirements, I devised a model for explanation intended to alleviate the limitations described previously. The model captures the explainer's reasoning about the design of an explanation and utilizes this knowledge to effectively support dialogue. The following are the main features of this model:

• Explanation knowledge is represented explicitly and separately from domain knowledge in a set of strategies that can be used to achieve the system's discourse goals.

• The system has many and varied strategies for achieving a given discourse goal.

• Utterances are planned in such a way that their intentional and rhetorical structure is explicit and can be reasoned about.

• The system keeps track of conversational context by remembering not only what the user asks, but also the planning process that led to an explanation.

• Information in the user model is utilized when it is available, but the system is able to operate effectively even when no pertinent information appears in the user model or when the user model is incorrect.

This model has been implemented in an explanation generation facility for the Explainable Expert Systems (EES) framework (Neches, Swartout, & Moore,

1985; Swartout et al., 1991), a domain-independent shell for creating expert system applications. When an expert system is built in EES, an extensive development history is created that records the domain goal structure and design decisions behind the expert system. This structure, as well as the system's static knowledge base, and the execution trace produced when the system is used to solve a particular problem, are all available for use by the explanation facility.

A detailed description of the implementation of the explanation facility is beyond the scope of this chapter, but may be found in Moore (in press) and Moore and Swartout (1989). Briefly, the explainer works in the following way: When the system needs to communicate with the user, a communicative goal (e.g., make the hearer know a certain concept, persuade the hearer to perform an action) representing an abstract specification of the response to be produced is posted. The system then uses its explanation strategies to plan an explanation to achieve the goal. When planning is complete, the system presents the explanation to the user, recording the *text plan* that produced it in a dialogue history. The text plan captures the "design" of an explanation, indicating what the system was trying to explain, how it explained it, and what alternative ways could have been used to explain the same thing. The dialogue history is a record of the conversation that has occurred thus far and includes the user's utterances as well as the text plans that lead to the system's responses. The text plans recorded in the dialogue history are used to provide the conversational context necessary to handle follow-up questions and recover when feedback from the user indicates that the system's explanation is not satisfactory.

We have found that the text plans that produced previous explanations are indispensable in determining how to interpret the hearer's feedback. Figure 4.4 shows a sample dialogue with the Program Enhancement Advisor (PEA), a prototype expert system implemented within the EES framework in order to test our explanation facility. PEA is an advice-giving system intended to aid users in improving their Common LISP programs by recommending transformations that enhance the user's code. This dialogue demonstrates that our system is able to participate in an ongoing dialogue with the user.

First, the two why-questions appearing on Lines 4 and 6 are interpreted differently because they appear in two different contexts. The first why-question occurs after the system has recommended that the hearer perform an action (Line 3). The system interprets this question as a request from users to be persuaded that they should perform this action. One of the strategies the system has for persuading hearers to perform actions is to state the shared goal (in this case "enhance maintainability") that led to its recommending the action, to state the method being applied to achieve this goal ("applying transformations that enhance maintainability"), and finally to state how the recommended act is involved in achieving the goal. This strategy leads to the system's response on Line 5. When the user asks "Why?" again after this response (Line 6), it is interpreted as a request for the system to justify the last statement it made in its explanation,

SYSTEM	What characteristics of the program would you like to enhance?	[1]
USER	Maintainability	[2]
	⋮	
SYSTEM	You should replace (LETQ x 1 with (SETF x 1).	[3]
USER	Why?	[4]
SYSTEM	I'm trying to enhance the maintainability of the program by applying transformations that enhance maintainability. SETQ-to-SETF is a transformation that enhances maintainability.	[5]
USER	Why?	[6]
SYSTEM	Transformations that enhance maintainability are defined as transformations whose right-hand-side's use is more general than its left-hand-side's use. SETF can be used to assign a value to any generalized-variable. SETQ can only be used to assign a value to a simple-variable.	[7]
USER	What is a generalized-variable?	[8]
SYSTEM	Simple-variables are storage locations named by symbols. This concept can be generalized to a storage location that can be named by any access function. These storage locations are called generalized-variables.	[9]
USER	Huh?	[10]
SYSTEM	For example, the *car* and *cdr* of a cons are generalized-variables, named by the access functions CAR and CDR. Other examples are an element of an array or a component of a structure.	[11]
USER	OK.	[12]

FIG. 4.4.   Sample dialogue with the PEA system.

leading to the interpretation "Why is SETQ-to-SETF a transformation that enhances maintainability?"

Another important aspect of this dialogue is that our system allows the user to ask the vaguely articulated question "Huh?" on Line 10. In order to answer the user's why-question on Line 6, the system must explain why SETQ-to-SETF is a transformation that enhances maintainability. In doing so, the system uses the term *generalized variable,* which is apparently unfamiliar to the user as evidenced by the follow-up question "What is a generalized variable?" on Line 8. In answering this question, the system uses one of its many strategies for describing a concept. In particular, it uses a strategy that reminds the user of a familiar concept (*simple variable*) which is a specialization of the concept being explained (*generalized variable*). The system then abstracts from the known concept to the new concept, focusing on the aspects (in this case, *named-by*) of the more specific concept that are being generalized to form the more general concept, which it then names. This is a very general strategy for describing a concept that can be applied whenever the user is familiar with a specialization of the concept to be described. In this case, the user does not understand this explanation, but cannot ask a pointed question elucidating exactly what is not understood. In our system, one of the options available to the user is to simply ask "Huh?", indicating that the previous explanation was not understood. The system "recovers" from this type of "failure" by finding another strategy for achieving the failed goal. In this case, the failed goal is to describe a concept, and the system recovers by giving examples of this concept, Line 11.

A more detailed discussion of how this dialogue is produced may be found in Moore (in press). We have also demonstrated that the text plans recorded in the dialogue history can be used to select the perspective from which to describe or compare objects (Moore, in press), as well as to avoid repeating information that has already been communicated (Moore & Paris, 1989). Building on this work, we are developing a system that uses its previous explanations to affect every subsequent explanation (Carenini & Moore, 1993; Rosenblum & Moore, 1993). We have also demonstrated (Moore & Swartout, 1990a) that the information in text plans allows the system to provide an intelligent hypertext-style interface in which users highlight the portion of the system's explanation they would like clarified, and the system produces a context-sensitive menu of follow-up questions that may be asked at the current point in the dialogue.

Discussion

As was the case with Method 1, comparing the behavior of an explanation system to that of human explainers has proven useful. Characterizing the differences and understanding the reasons for the disparities allows us to identify specific aspects of explanation systems that must be improved.

One issue that must be addressed when using Method 2 for assessing explanation systems is how far to carry the analogy between human–human interactions and human–machine interactions. Certainly there are differences between the capabilities of humans and machines, and we should not blithely assume that human–computer interaction must mimic human–human interaction in all aspects. Rather, we must make reasoned decisions about what aspects of human–human interaction we wish to preserve in human–machine interaction, what differences can be tolerated, and in what ways human–computer interaction can improve upon human–human interaction.

In some cases, intelligent systems may offer benefits beyond the scope of human explainers. For example, expert systems differ from human experts in that, if they are built according to certain principles, expert systems have access to their problem-solving strategies and can accurately report exactly what methods were used to solve a problem and why these methods were chosen (see for example Neches et al., 1985; Swartout et al., 1991). Human experts, on the other hand, do not have access to the actual methods they employ in reasoning. They can, however, construct a justification for why a solution is correct or reconstruct a plausible chain of reasoning based on their rich model of the domain. Within the expert system explanation community, there is currently a debate about how closely the "line of explanation" must follow the system's "line of reasoning." Wick and Thompson (1992) argued that an effective explanation need not be based on the actual reasoning processes that the system used in solving the problem, but rather, the system's results may be supported by other sources of information about the domain. However, Swartout argued that such an approach is at odds with one of the desired attributes for expert system explanation, namely *fidelity*, which requires that the explanation be an accurate representation of what the expert system really does. Regardless of how this issue is resolved, the fact that expert systems may have access to their line of reasoning affords an opportunity not available to human experts.

The hypertext-style interface we designed for our explanation facility provides another example of the way in which human–computer interaction may usefully differ from human–human interaction. We designed this interface to provide users with a convenient way to ask questions about previously given explanations. In human–human interactions, people can ask questions that refer to previous explanations with utterances such as "I didn't understand that part about applying transformations that enhance maintainability?" But such questions pose a difficult challenge for natural language understanding because such questions often intermix metalevel references to the discourse with object-level references to the domain. Our hypertextlike interface allows users to point to the portion of the system's explanation that they would like clarified. For example, suppose the user wished to ask a follow-up question about applying transformations that enhance readability after turn 5 in the sample dialogue in Fig. 4.4. To get a menu of follow-up questions, the user would employ the mouse to highlight the string "applying transformations that enhance maintainability" where it appears on the screen. By allowing users to point, many of the difficult referential problems in natural language analysis can be avoided. Implementing this interface was possible in our explanation facility precisely because our system understands both the structure and content of its own explanations, and thus is able to understand what the user is pointing at. The system then offers the user a menu of follow-up questions that it knows how to answer and that make sense in the current context. Although this interface differs from what occurs in human–human interaction, it provides a pragmatic solution to the problem of allowing users to ask questions about previous explanations. It may even be the case that when interacting with a computer, users prefer to highlight the text they would like to ask about and to receive a menu of possible questions, rather than attempting to formulate a natural language question. We plan to test this hypothesis in future work.

METHOD 3: CRITERIA FOR EVALUATING
EXPLANATION SYSTEMS

The third method for assessing explanation systems is to devise a set of criteria for evaluation and to rate systems according to these criteria. Swartout (1990) attempted to codify a set of desiderata for explanation facilities. He suggested that these requirements can be used as metrics for evaluating the performance of explanation systems and progress in the field. The desiderata, shown in Fig. 4.5, fall into three classes. The first places constraints on the mechanism by which explanations are produced. The second and third specify requirements on the explanations themselves. The fourth and fifth are concerned with the effects of an explanation facility on the construction and execution of the expert system of which it is a component.

---

1. **Fidelity.** Explanations must be an accurate representation of what the expert system really does.

2. **Understandability.** Explanations must be understood by users. Understandability is not a single factor, but is made up of several factors, including:

   Terminology. The terms used in explanations must be familiar to the user or the system must have the capability to define them.
   Abstraction. The system must be able to give explanations at different levels of abstraction of terminology. For example, in describing a patient's problem, the system should be able to use the abstract term bacterial infection, or the more specific term *E. coli* infection.
   Summarization. The system must be able to provide descriptions at different levels of detail.
   Viewpoints. The system must be able to present explanations from different points of view that take into account the user's interests and goals.
   Linguistic Competence. The system should produce explanations that sound "natural."
   Coherence. Taken as a whole, the explanation should form a coherent set. Explanations should take into account previous explanations.
   Compatability. When several things must be conveyed in a single explanation, there shuld be smooth transitions between topics.
   Correct Misunderstandings. The system must allow the user to indicate that an explanation is unsatisfactory and be capab le of providing further clarification.

3. **Sufficiency.** Enough knowledge must be represented in the system to support production of the kinds of explanations that are needed. The system must be able to handle the types of questions that users wish to ask.

4. **Low Construction Overhead.** Providing explanation should impose light load on expert system construction, or any load that is imposed should be recovered by easing some aspect of the expert system lifecycle (e.g., maintenance of evolution).

5. **Efficiency.** The explanation facility should not degrade the routine efficiency of the expert system.

FIG. 4.5. Swartout's desiderata for explanation systems.

Swartout went on to describe the implications of some of these criteria. For example, the need for fidelity has several implications. First, the explanations must be based on the same underlying knowledge that the system uses for problem solving. Thus systems that produce explanations using canned text or fill-in-the-blank templates would receive a poor rating because there can be no guarantee that explanations produced by these systems are consistent with the program's behavior. Another implication of the fidelity criterion is that the expert system's inference engine should be as simple as possible with a minimal number of special features built into the interpreter. Such features are not part of the explicit knowledge base of the system, and either are not explained at all or are explained by special-purpose routines built into the explanation facility. This introduces the potential for inaccuracy because changes made to the interpreter require that changes be made independently to the explanation routines. For example, the certainty factor mechanism that manages MYCIN's reasoning about uncertainty cannot be explained because it is built into MYCIN's interpreter.

The efficiency metric has important implications as well. It would rate a system that provided good justifications of its actions as poor if that system did so by always reasoning from first principles. Such a system would be re-deriving its expertise on each run and would be very inefficient.

Discussion

The criteria proposed by Swartout are very comprehensive and are quite useful as qualitative guidelines. However, it would be desirable to form evaluation metrics from these criteria with objective methods for assigning ratings to an explanation system. In some cases, the task of devising a method for assigning a value to the metric seems straightforward. For example, fidelity can be assessed by comparing traces of the system's problem solving with the natural language explanations it produces. Techniques from software engineering could be helpful in estimating the overhead in system construction due to the explanation facility and how much savings in the maintenance and evolution cycles are due to design decisions attributable to the requirements imposed by explanation. Cases in which systems have been reengineered to provide better explanation capabilities also offer opportunities for evaluation. For example, because NEOMYCIN was developed to address certain of the limitations in MYCIN's explanation capabilities, we can get an estimate of runtime efficiency by comparing NEOMYCIN to MYCIN solving the same diagnostic problem. Moreover, the experiences of knowledge engineers working on both systems should give insight into whether the restructuring of the knowledge base for NEOMYCIN aids in maintenance or evolution.

Some aspects of understandability could also be objectively measured. For example, one way to evaluate the factor of composability (smoothness between topic transitions in a single explanation) would be to analyze the system's explanations to determine whether they adhere to constraints governing how focus of attention shifts, as defined by Sidner (1979) and extended by McKeown (1985).

In other cases, it is difficult to envisage how objective measures for assessment could be devised. For example, how can we assign a value to an explanation's naturalness (linguistic competence) and coherence? The ratings of such factors are inevitably subjective and can only be judged by human users. Furthermore, what is understandable to one user may be obscure to others. The most promising way to assess the understandability of a system's explanation will involve techniques such as those included in the discussion of Method 1, that is, assessing the users' satisfaction with the explanations or the impact of the explanations on users' performance.

The criteria proposed by Swartout provide a good starting point for devising metrics for assessment, but clearly much more work needs to be done. In the next section, two examples of improved architectures for explanation are discussed.

EXPLAINABLE ARCHITECTURES FOR EXPERT SYSTEMS

The insights gained from analyses of the inadequacies of early expert systems led researchers to attempt to design architectures for expert systems that would improve their explanation capabilities. In designing new architectures, researchers had the goals of capturing the knowledge needed to support the types of explanation users desire, and to structure that knowledge approximately. Here I briefly discuss two such systems.

NEOMYCIN

NEOMYCIN (Clancey & Letsinger, 1984) was developed in order to teach medical students about diagnosis, and for this purpose it was necessary to be able to justify the diagnostic associations encoded in MYCIN's rules and to explicate the overall diagnostic strategy of gathering information and focusing on hypotheses. Recall that MYCIN could not answer questions of these types due to limitations in its knowledge base. NEOMYCIN was designed with the goal of capturing control knowledge more explicitly so that it could be explained and re-used. Clancey argued that NEOMYCIN's metarules constitute a domain-independent diagnostic strategy that could be applied to related problems in other domains (Clancey, 1983a).

In NEOMYCIN, a domain-independent diagnostic strategy is represented explicitly and separately from knowledge about the domain (the disease taxonomy, causal and data/hypothesis rules, and world facts). To build an expert system using NEOMYCIN, the developer must first identify the "task" structure of the problem, for example, *make-diagnosis, pursue-hypothesis, explore-and-*

---

17) Has Mary been hospitalized recently?

**WHY?

⋮

**HOW 25.0?

[i.e., HOW are we trying to develop the differential using basic history and pertinent physical exam?]

We did the following:

[26.1] Got a general idea of the problem: categorized it into one of the several pathogenic classes or disease loci, or both.

[26.2] Confirmed and refined the differential diagnosis through specific questions.

We are now:

[26.3] Rounding out the diagnostic information by looking generally into past medical history and by reviewing systems.

[back to question 17 ....]

**NO

FIG. 4.6.   A sample NEOMYCIN explanation.

refine. A diagnostic strategy is then represented as a set of *tasks,* which are meta-level goals, and *metarules* (Davis, 1980) for achieving these goals. An ordered collection of metarules defines as *generic procedure* for achieving a task. Next, domain-specific rules are organized into rule sets based on this task structure. Rule sets become active depending on which tasks have been posted.

Because NEOMYCIN's strategic knowledge is explicitly represented, the system can produce explanations of its problem-solving strategies. For example, Fig. 4.6 (from Hasling et al., 1984) shows that NEOMYCIN is able to give abstract explanations of its general problem-solving strategy.[3] In addition, the possibility of knowledge re-use exists if indeed NEOMYCIN's metarules constitute a domain-independent diagnostic strategy.

However, earlier we saw an example of NEOMYCIN's explanation behavior that indicated the improved representation of strategic knowledge alone was not sufficient to improve all types of explanations (see Fig. 4.3). This example showed that more work must be done to improve the understandability of the system's explanations. In particular, techniques for summarization and choosing the appropriate level of abstraction for an explanation in a particular situation must be devised. The efficiency of the system is also an issue that must be

---

[3]In this figure, user input follows the double asterisks and is depicted in capital letters (e.g., "**WHY?"). The system numbers each question it asks as well as each clause of the explanations that it produces. The user can ask questions about a portion of a previous explanation by referring to its number (e.g., "How 25.0?")

evaluated, because NEOMYCIN essentially rediscovers its expertise each time it solves a problem.

Explainable Expert Systems (EES) Framework

EES (Neches et al., 1985; Swartout et al., 1991), like its precursor XPLAIN (Swartout, 1983), grew out of the observation that much of the knowledge needed to produce explanations is not explicitly represented in the expert system's knowledge base, but instead remains in the system builder's head. The goal of EES was to create a framework, or "shell," to capture this knowledge. The approach taken in designing the EES framework was to first identify the types of explanations that must be produced. This, in turn, indicated the kinds of knowledge that must be represented. The EES language allows a system developer to provide knowledge about an application domain, mainly in terms of an abstract specification of how the domain works, how problems are solved in the domain, and domain terminology. The system builder is then replaced by an automatic programer that synthesizes the desired expert system from this specification, recording all the design decisions made during the program-writing process. The explanation routines then use the recorded design history to provide richer explanations, including justications for actions.

A sample of the explanations produced by the PEA system, which was built within the EES framework, was shown in Fig. 4.4. These explanation capabilities were possible for two reasons. First, the EES framework provides the types of knowledge needed to support explanations. Second, explanation is treated as a sophisticated problem-solving activity requiring its own knowledge and expertise. Techniques from natural language generation and new techniques for dialogue management were incorporated into the explanation facility (Moore, in press). Other expert systems are being developed using EES (Paris, 1991a) and research on the explanation facility is continuing.

In addition to its implications for explanation, we have found that the EES approach offers other advantages related to development and maintenance. For example, the discipline imposed by the knowledge representation in EES provides guidance for knowledge engineers during the development process. We believe this rigor will make errors and inconsistencies in the knowledge base easier to detect. In addition, because the automatic program writer creates an executable expert system, it is the compiled code that is actually executed, but the rationale for that code is available for explanation. Thus representing the knowledge needed for explanation does not incur runtime overhead; the system is not re-deriving its expertise on every run. In terms of construction overhead, it is clearly more work to develop an expert system using EES because more knowledge must be represented than in a system such as MYCIN. However, we have found that maintenance and evolution are facilitated because modifications are performed at the knowledge base (i.e., "specification") level, rather than at the implementation level. Addition of a new domain concept requires making a few assertions and rerunning the automatic program writer rather than extensive manual recoding.

In addition, as in NEOMYCIN, we believe that separating different forms of knowledge will also reduce the amount of work that has to be done to move to a new domain. EES has been used to construct systems in several domains: an advice-giving system that aids users in enhancing their LISP programs, a diagnostic system that locates faults in simple electronic circuits, and a diagnostic system for identifying faults in a local area network. While we are gaining empirical evidence that the problem-solving architecture of EES is general enough to support several classes of problems, in particular advice-giving and diagnosis tasks, we are also obtaining information about the generality of the explanation facility we proposed in EES. In particular, we are gaining experience in determining how viable it is to provide a domain-independent explanation component.

CURRENT DIRECTIONS

The studies of conventional expert systems explanation facilities discussed previously led to several important observations. First, inadequacies in the knowledge bases of early systems were identified and led to knowledge bases that separately and explicitly represent the types of knowledge needed to support explanations. These included justifications of the systems' actions, explications of general problem-solving strategies, and definitions of terminology. Second, we have realized that explanation is a problem in its own right, requiring its own expertise and a sophisticated problem-solving architecture. The improvements that came simply by improving expert system knowledge bases were not sufficient. In fact, the knowledge bases of newer expert systems that separate different types of knowledge (Swartout et al., 1991) and represent knowledge at various levels of abstraction (Patil, Szolovits, & Schwartz, 1984), confront explanation generators with any array of choices about what information to include in an explanation and how to present that information that were unavailable in the simple knowledge bases of earlier systems. Meeting the challenge posed by these richer knowledge bases will require more sophisticated and linguistically motivated explanation generators. Further, we now understand that explanation cannot be an afterthought; it must be designed into the system from the outset.

We have seen the emergence of explainable expert system frameworks, such as EES and (possibly) NEOMYCIN, that provide system builders with the tools they need to develop systems that will be explainable. Experience with these shells has shown that building an explainable system requires more work during system development, but that the discipline imposed by explanation requirements improves the architecture for other aspects of the software life cycle, in particular, knowledge acquisition and system maintenance.

The most promising course for the future is to provide system developers with an explainable intelligent system shell that can be customized for specific application domains. The shell would provide a domain-independent knowledge base, weak methods for problem solving, domain-independent explanation strategies, a lexicon for closed-class words, and user model acquisition facilities. This shell could also be augmented with tools for adding domain-specific knowledge, such as editors, authoring tools, browsing tools, and domain lexicons.

Whereas the feasibility of this approach can only be verified empirically as more systems are developed using shells such as the EES framework, the community has learned much that can be useful in the assessment of explanation systems by identifying the constraints that the need to provide explanations places on the knowledge representation and reasoning processes of an intelligent system. These requirements themselves provide a set of metrics that can be used to determine whether a system can readily accept an explanation module.

ACKNOWLEDGMENTS

Preparation of this chapter was supported in part by the Office of Naval Research Cognitive and Neural Sciences Division, the National Science Foundation, the National Library of Medicine, and a contract from the Advanced Research Projects Agency, administered by the Office of Naval Research (ONR), to the UCLA Center for the Study of Evaluation/Center for Technology Assessment. However, the opinions expressed do not necessarily reflect the positions of ONR, NSF, ARPA, or the MLM, and no official endorsement by any of these organizations should be inferred. The research on the Explainable Expert Systems framework described here was supported by the Advanced Research Projects Agency under grant MDA 903-81-C-0335 and under NASA Ames cooperative agreement NCC2-520. I gratefully acknowledge Sean McLinden and Alan Lesgold for helpful comments on earlier drafts of this chapter.

REFERENCES

Alessi, S. M., & Trollip, S. R. (1991). *Computer-based instruction: Methods and development.* Englewood Cliffs, NJ: Prentice-Hall.

Appelt, D. E. (1985). *Planning English sentences.* Cambridge: Cambridge University Press.

Buchanan, B. G., & Shortliffe, E. H. (1984). *Rule-based expert systems: The MYCIN experiments of the Stanford Heuristic Programming Project.* Reading, MA: Addison-Wesley.

Carenini, G., & Moore, J. D. (1993). Generating explanations in context. *Proceedings of the International Workshop on Intelligent User Interfaces.* W. D. Gray, W. E. Hefley, & D. Murray (Eds.). ACM Press, pp. 175–182.

Cawsey, A. (1993). Planning interactive explanations. *International Journal of Man–Machine Studies 38*(2), 169–200.

Chin, D. N. (1989). Knome: Modeling what the user knows in uc. In A. Kobsa & W. Wahlster

(Eds.), *User models in dialog systems* (pp. 74–107). New York: Springer-Verlag, Symbolic Computation Series.

Clancey, W. J. (1983a). The advantages of abstract control knowledge in expert system design. *Proceedings of the Third National Conference on Artificial Intelligence*, 74–78.

Clancey, W. J. (1983b). The epistemology of a rule-based expert system: A framework for explanation. *Artificial Intelligence, 20*(3), 215–251.

Clancey, W. J., & Letsinger, R. (1984). Neomycin: Reconfiguring a rule-based expert system for application to teaching. In *Readings in medical artificial intelligence: The first decade* (pp. 361–381). Addison-Wesley.

Davis, R. (1980). Meta-rules: Reasoning about control. *Artificial Intelligence, 15,* 179–222.

Davis, R., & Lenat, D. B. (1982). *Knowledge-based systems in artificial intelligence.* New York: McGraw-Hill.

Finin, T. W., Joshi, A. K., & Webber, B. L. (1986). Natural language interactions with artificial experts. *Proceedings of the IEEE, 74*(7).

Fischer, G. (1987). A critic for LISP. *Proceedings of the 10th International Joint Conference on Artificial Intelligence,* 177–184. Los Altos, CA: Morgan Kaufmann.

Hasling, D. W., Clancey, W. J., & Rennels, G. (1984). Strategic explanations for a diagnostic consultation system. *International Journal of Man–Machine Studies, 20*(1), 3–19.

Kass, R. (1991). Building a user model. *User Model and User Adapted Interaction, 1*(3), 203–258.

Kobsa, A. (1989). A taxonomy of beliefs and goals for user models in dialog systems. In A. Kobsa & W. Wahlster (Eds.), *User models in dialog systems* (pp. 52–68). New York: Springer-Verlag, Symbolic Computation Series.

Littman, D., & Soloway, E. (1988). Evaluating ITSs: The cognitive science perspective. In M. C. Polson & J. J. Richardon (Eds.), *Foundations of Intelligent Tutoring Systems* (pp. 209–242). Hillsdale, NJ: Lawrence Erlbaum Associates.

Mastaglio, T. W. (1990). *User modelling in cooperative knowledge-based systems.* Unpublished doctoral dissertation, Department of Computer Science, University of Colorado, Boulder.

McCoy, K. F. (1989). Generating context sensitive responses to object-related misconceptions. *Artificial Intelligence, 41*(2), 157–195.

McKeown, K. R. (1985). *Text generation: Using discourse strategies and focus constraints to generate natural language text.* Cambridge: Cambridge University Press.

McKeown, K. R. (1988). Generating goal-oriented explanations. *International Journal of Expert Systems, 1*(4), 377–395.

Moore, J. D. (in press). *Participating in explanatory dialogues: Interpreting and responding to questions in context.* Cambridge, MA: MIT press.

Moore, J. D. (1989). Responding to "huh?": Answering vaguely articulated follow-up questions. *Proceedings of the Conference on Human Factors in Computing Systems,* 91–96.

Moore, J. D., & Paris, C. L. (1989). Planning text for advisory dialogues. *Proceedings of the 27th Annual Meeting of the Association for Computational Linguistics,* 203–211.

Moore, J. D., & Swartout, W. R. (1989). A reactive approach to explanation. *Proceedings of the 11th International Joint Conference on Artificial Intelligence,* 1504–1510.

Moore, J. D., & Swartout, W. R. (1990). Pointing: A way toward explanation dialogue. *Proceedings of the National Conference on Artificial Intelligence,* 457–464.

Moore, J. D., & Swartout, W. R. (1991). A reactive approach to explanation: Taking the user's feedback into account. In C. L. Paris, W. R. Swartout, & C. W. Mann (Eds.), *Natural language generation in artificial intelligence and computational linguistics* (pp. 3–48). Boston: Kluwer Academic Publishers.

Neches, R., Swartout, W. R., & Moore, J. D. (1985). Enhanced maintenance and explanation of expert systems through explicit models of their development. *IEEE Transactions on Software Engineering, SE-11*(11), 1337–1351.

Paris, C. L. (1991a). Generation and explanation: Building an explanation facility for the explainable expert systems framework. In C. L. Paris, W. R. Swartout, & C. W. Mann (Eds.), *Natural language generation in artificial intelligence and computational linguistics* (pp. 49–81). Boston: Kluwer Academic Publishers.

Paris, C. L. (1991b). The role of the user's domain knowledge in generation. *Computational Intelligence, 7*(2), 71–93.

Patil, R. S., Szolovits, P., & Schwartz, W. B. (1984). Causal understanding of patient illness in medical diagnosis. In W. T. Clancey & E. H. Shortliffe (Eds.), *Readings in medical artificial intelligence: The first decade* (pp. 339–360). Reading, MA: Addison-Wesley.

Pavlin, J., & Corkill, D. D. (1984). Selective abstraction of AI system activity. *Proceedings of the National Conference on Artificial Intelligence,* 264–268. Menlo Park, CA: AAAI.

Pollack, M. E., Hirschberg, J., & Webber, B. L. (1982). User participation in the reasoning processes of expert systems. *Proceedings of the Second National Conference on Artificial Intelligence,* Menlo Park, CA: AAAI.

Reichman-Adar, R. (1984). Extended person–machine interface. *Artificial Intelligence, 22*(2), 157–218.

Rich, E. (1989). Stereotypes and user modelling. In A. Kobsa & W. Wahlster (Eds.), *User models in dialog systems* (pp. 35–51). New York: Springer-Verlag, Symbolic Computation Series.

Ringle, M. H., & Bruce, B. C. (1981). Conversation failure. In W. G. Lehnert & M. H. Ringle (Eds.), *Knowledge representation and natural language processing* (pp. 203–221). Hillsdale, NJ: Lawrence Erlbaum Associates.

Robinson, J. J. (1984). *Extending grammars to new domains* (Technical Report No. ISI/RR-83-123). Los Angeles, CA: USC/Information Sciences Institute.

Rosenblum, J. A., & Moore, J. D. (1993, August). Participation in instructive dialogues: Finding and exploiting relevant prior explanations. *Proceedings of the World Conference on Artificial Intelligence in Education,* pp. 145–152. Association for the Advancement of Computing in Education, Charlottesville, VA.

Rubinoff, R. (1985). *Explaining concepts in expert systems: The CLEAR system* (Tech. Rep. No. MS-CIS-85-06). Philadelphia: University of Pennsylvania.

Shortliffe, E. H. (1976). *Computer based medical consultations: MYCIN.* New York: Elsevier North Holland.

Sidner, C. L. (1979). *Toward a computational theory of definite anaphora comprehension in English discourse.* Unpublished doctoral dissertation, MIT, Cambridge, MA.

Sparck Jones, K. (1989). Realism about user modelling. In A. Kobsa & W. Wahlster (Eds.), *User models in dialog systems* (pp. 341–363). New York: Springer-Verlag, Symbolic Computation Series.

Suchman, L. A. (1987). *Plans and situated actions: The problem of human–machine communication.* Cambridge: Cambridge University Press.

Suthers, D. D. (1991). Task-appropriate hybrid architectures for explanation. *Computational Intelligence, 7*(4).

Swartout, W. R. (1983). XPLAIN: A system for creating and explaining expert consulting systems. *Artificial Intelligence, 21*(3), 285–325.

Swartout, W. R. (1990, July). *Evaluation criteria for expert system explanation.* Paper presented at the AAAI90 Workshop on Evaluation of Natural Language Generation Systems, Boston.

Swartout, W. R., & Moore, J. D. (1993). Explanation in second generation expert systems. In J. M. David, J. P. Krivine, & R. Simmons (Eds.), *Second Generation Expert Systems.* New York: Springer-Verlag.

Swartout, W. R., Paris, C. L., & Moore, J. D. (1991). Design for explainable expert systems. *IEEE Expert, 6*(3), 58–64.

Teach, R. L., & Shortliffe, E. H. (1984). An analysis of physicians' attitudes. In B. G. Buchanan & E. H. Shortliffe (Eds.), *Rule-based expert systems: The MYCIN experiments of the Stanford Heuristic Programming Project,* (chap. 34, pp. 635–652). Reading, MA: Addison-Wesley.

van Beek, P. (1987). A model for generating better explanations. *Proceedings of the 25th Annual Meeting of the ACL.*

Wallis, J. W., & Shortliffe, E. H. (1984). Customized explanations using causal knowledge. In *Rule-based expert systems: The MYCIN Experiments of the Stanford Heuristic Programming Project* (chap. 20, pp. 371–388). Addison-Wesley.

Webber, B. L., & Joshi, A. (1982). *Taking the initiative in natural language data base interactions: Justifying why* (Tech. Rep. No. MS-CIS-82-1). Philadelphia: University of Pennsylvania.

Weiner, J. L. (1980). BLAH, a system which explains its reasoning. *Artificial Intelligence, 15:*19–48.

Wick, M., & Thompson, W. B. (1992). Reconstructive expert system explanation. *Artificial Intelligence, 54,* 33–70.

Wolz, U. (1990). The impact of user modeling on text generation in task-oriented settings. *Proceedings of the Second International Workshop on User Modeling.*

Wu, D. (1991). Active acquisition of user models: Implications for decision-theoretic dialog planning and plan recognition. *Journal of User Model and User Adapted Interaction, 1*(2), 149–172.

Young, E. A. (1987). *Using the full power of computers to learn the full power of computers.* (Tech. Rep. No. 87-9). Boulder, CO: Institute of Cognitive Science, University of Colorado.

5 Assessment of Enabling Technologies for Computer-Aided Concurrent Engineering (CACE)

Azad M. Madni
Perceptronics, Inc., Woodland Hills, CA

In today's competitive industrial environment, a major priority is the development of new and novel approaches for dramatically reducing product development times while improving overall product quality. The concept that has become pivotal to achieving these objectives is concurrent engineering (CE). Concurrent engineering offers several advantages over traditional sequential engineering, including shorter product development times, superior product quality, dramatically higher user acceptance, lower cost, and higher assurance of meeting time-to-market requirements. Concurrent engineering calls for early and regular collaboration among engineering, manufacturing, management, and support personnel during the product planning and design processes.

This chapter provides an assessment of enabling technologies underlying concurrent engineering. It presents a conceptual framework that provides the basis for discussing the technological components. These include a collaborative design environment, "executable" process models and design specifications, a formal approach to human–machine integration, interactive multimedia technologies, and computer-aided concurrent engineering tools and their integration.

In the last three decades computer-aided automation has seen a steady increase in the different phases and facets of a product's life. The factory of the future seemed imminent. But after huge expenditures and frequent speculation, it became apparent that the design automation concepts and sophisticated equipment could not keep up with an environment plagued by constant change. Variations in raw material and equipment breakdown are but a few examples of how "change" is the rule, not the exception, in manufacturing environments. Moreover, despite the manifest advantages of "soft" automation, there are still excessive delays in time-to-market that, in part, negate any advantages.

The fundamental problem is that the traditional approach to solving large, complex engineering problems is inherently sequential; that is, the problem is decomposed into its constituent subproblems and the subproblems are tackled sequentially—moving from research and development of materials and processes, to product design, manufacture, installation, and support. This approach has obvious drawbacks including late discovery of problems; long, costly design iterations; suboptimal solutions due to insufficient evaluation of options early in design; long product development times; concomitant negative impacts of product cost, quality, supportability; and last minute engineering changes due to design shortcomings discovered at manufacturing time.

In light of these manifest deficiencies with sequential engineering, the industry is turning to *concurrent engineering*. Winner, Pennell, Bertrand, and Slusarczuk (1988) offered the following definition of concurrent engineering:

> Concurrent Engineering is a systematic approach to the integrated concurrent design of products and their related processes, including manufacture and support. This approach is intended to cause the developers, from the outset, to consider all elements of the product life cycle from conception through disposal, including quality, cost, schedule, and user requirements. (p. 2)

This chapter presents a conceptual framework for concurrent engineering, along with an assessment of enabling technologies and tools that can promote and naturally enforce concurrent engineering principles and practices.

COMPUTER-AIDED CONCURRENT ENGINEERING

In a concurrent engineering environment, multidisciplinary teams consisting of different members from the design, manufacturing, and support functions work together with the customer on all phases of product development, fully sharing information and participating in decision making (Winner et al., 1988).

The challenge of concurrent engineering is to overcome organizational fragmentation and manage complexity through a combination of cultural changes and technological innovation. Concurrent engineering requires a new approach to accountability, focus, and coordination of multiple objectives-oriented teams. Specifically, the introduction of concurrent engineering requires that both vision and knowledge must be aligned across the different functional groups (materials, engineering, manufacturing, management, and support services).

The concurrent engineering team usually works under a single budget. Because the team has a common goal, design changes that provide an overall benefit to the team can be identified quickly. Resolution of design deficiencies does not have to be deferred, but can be addressed as soon as deficiencies are discovered. This strategy is key to producing a robust system that incorporates the best or acceptable features for all.

With respect to the insertion of concurrent engineering methods and tools within design and manufacturing environments, the challenges are managing application complexity while producing a useful product; managing cultural change while introducing concurrent engineering principles and practices; managing technical and technological difficulty without losing focus; leveraging existing tools and inplace procedures, to the extent possible, without violating concurrent engineering principles; and measuring incremental progress to ascertain accomplishment of interim milestones without interfering with ongoing work.

Whereas concurrent engineering has been accepted in concept by the engineering and manufacturing communities, a unified framework for introducing concurrent engineering practices, procedures, and tools is necessary to feel the full impact of concurrent engineering. The concurrent engineering process (Fig. 5.1) is conceptualized from this viewpoint. The main idea behind this process is to allow the designers to look "down the line" while still in the early stages of the design process. Figure 5.1 shows the overall process and sequence of design steps undertaken by the collaborative design teams in the concurrent design of a product.

Figure 5.1 illustrates a conceptual framework for concurrent engineering (Madni, Balcerak, Estrin, Freedy, & Melkanoff, 1990). Starting with a preliminary set of requirements and initial product specifications, the design teams collaborate in the design process within a design environment that supports electronic mail, teleconferencing, and sharing of all "design objects" (e.g., partial solutions, design versions, and tools). The *preliminary design* process is facilitated by "rough cut" process modeling and analysis tools. The design teams construct high-level process models in the application domain with the help of the process modeling tools. The preliminary design is progressively *refined* into detailed process models for in-depth analysis. When the process and product design specification becomes relatively "stable," the manufacturing process is simulated and evaluated prior to making a commitment to hardware. The *process simulation* requires "executable" process models for simulating the different flows (e.g., parts, tools, people, dollars, and information) in a manufacturing enterprise. Several design versions are created during the course of this simulate-and-evaluate cycle. These are catalogued in the order created along with attendant assumptions, decisions, and constraints, thereby creating a design history and design version audit trail for future use by the design teams.

ENABLING TECHNOLOGIES

Realizing the full benefits of concurrent engineering is far more than a technological problem. In fact, to realize the full impact of concurrent engineering requires a fundamental cultural change at all levels in an organization. The technology

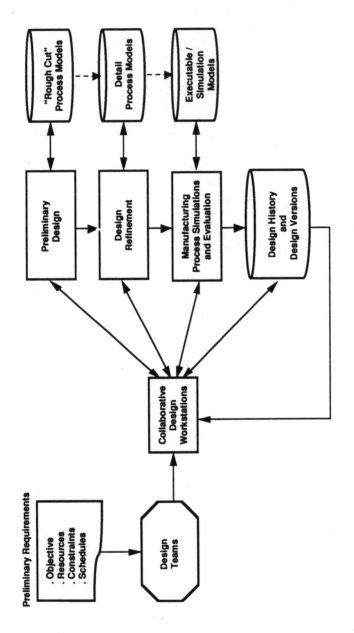

FIG. 5.1. Concurrent engineering paradigm. From "Computer-Aided Concurrent Engineering (CACE) of Infrared Focal Plane Arrays (IRFPAs): Emerging Directions and Future Prospects" by A. M. Madni, R. Balcerak, G. Estrin, A. Freedy, and M. A. Melkanoff (1990). Paper presented at the Second National Symposium on Concurrent Engineering. Reprinted by permission.

assessment described in this chapter is from two perspectives: (a) how technology can enhance the simultaneous development of the product and the process, and (b) how technology can provide the basis for realizing a much-needed cultural change. Within the overall framework of Fig. 5.1 we discuss and assess five different enabling technologies: (a) collaborative design environments, (b) "executable" process models and design specifications, (c) human–machine integration methodologies, (d) interactive multimedia technologies, and (e) computer-aided concurrent engineering tools and tool integration techniques.

Computer-Aided Collaborative Design

A key element of the design process is to ensure that all key decision-makers and system operators can participate on design teams. Because it is not usually practical to have collaborative sessions involving all technical disciplines (e.g., manufacturing operations manager, design engineers, production engineers, reliability/maintainability engineers, systems designer, and shop floor managers), the project leader or task force needs to partition tasks to be carried out by teams of four to six people and to plan communication and coordination needed between the design teams. To expand communication, each member of the design team needs access to and protocols for the use of terminals and servers that provide full, high resolution interactive graphics capabilities. The different terminals, workstations, or mainframes need to be interconnected via a local area/wide area network using X-Window protocols as the graphics communication medium. The members of any design team need to be supported by an environment that helps them communicate easily with each other, to coordinate their activities, and to share common design objects (e.g., design representations, design history, and design tools).

Figure 5.2 shows a high-level view of the collaborative design environment. A large-screen display facilitates focus on the issues being discussed whenever a subgroup meets face to face in the same room. In those situations where subgroup participants in a design session are remote from each other, inherent delays make synchronous interaction more difficult, so that more structured protocols (e.g., topics of discourse templates) are necessary to guide orderly discussion and decision making. The open systems that have become available recently involving heterogeneous servers and workstations using UNIX™, Ethernet, and X-Windows have made possible several interactive functions that were previously infeasible such as distributed data sharing, dynamic visualization of scientific data, and collaborative work. The process can be helped by teleconferencing technology, including FAX and video hookups. Of crucial importance to this work is the extent to which each participant can be made aware of changes to design objects and the ease with which each can access the state of such objects

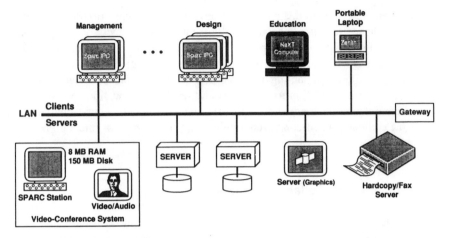

FIG. 5.2. Networked client-server collaborative design environment architecture.

in a global data base. Figure 5.3 shows the functionalities required from the collaborative design environment.

Although collaborative design is a well-received concept, there are still some technical hurdles (e.g., the ability to share design "objects") that have to be overcome to develop an effective collaborative design environment (Mujica et al., 1990).

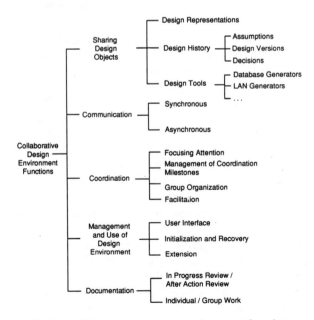

FIG. 5.3. Collaborative design environment functions.

Executable Process Models

Process models allow description, integration, and evaluation of the different "flows" in a manufacturing enterprise at different levels of detail and from multiple perspectives (Madni, 1988a; Madni et al., 1990; Estrin, Fenchel, Razouk, & Vernon, 1986).

Process models, or models that produce executable specifications (Harel et al., 1988; Madni et al., 1990), allows the design team to analyze the impact of "downstream" constraints on candidate designs with a view to achieving a design that satisfies manufacturing, assembly, cost, and support constraints. Table 5.1 provides a summary of the desired characteristics of process modeling and simulation.

Human–Machine Integration

In the foreseeable future, humans will continue to serve as "enabling components" in a manufacturing enterprise. Proper integration of humans and equipment can make all the difference between a relatively trouble-free factory and one that is beleaguered with human errors arising from a poor integration of humans and machines. The approach of Madni et al. (1990) to human–machine integration relies on four different classes of simulation. This family of simulations is designed to produce the most cost-effective solutions at various stages of the concept development/exploration and demonstration/validation phases of the

TABLE 5.1
Desired Characteristics of Process Modeling and Simulation

Accept several kinds of inputs for factory simulation, including graphics, high-level language declarations, and program modules written in specific languages.
Combine visual formalisms with executable specifications.
Provide default description of various processes/subprocesses.
Permit explicit description of the key flows and parameters:

People		
Authority	]	allows description of the organizational model
Parts		
Tools		
Information		
Cost	]	basis for a managerial model

Allow people/models to be included at various levels within the simulation.
Automatically generate data bases, local area networks, and process control software.
Allow systematic application of "downstream" constraints in a "what-if" design process.
Allow progressive top-down refinement of the simulation (higher and lower levels) and support bottom-up integration (from lower to higher levels).
Integrate with the rest of the enterprise, thus allowing

The concurrent design team to test the manufacturability and cost-effectiveness of the design.
The factory personnel to integrate the different tools in their ongoing operations.

TABLE 5.2
The Simulation Continuum From a Human-Machine Integration Perspective

Simulation Type	Analytic Simulation	Interactive Simulation	Reconfigurable Physical Simulator	Networked Simulation
Comparison Criteria				
Objectives	Function, allocation, and workload analysis Information analysis	Operator error and error rates assessment Concept of operation specification/ refinement	Operability assessment Control display layout optimization	Communication load and error analyses Team training
Human component	(e.g., rule-based)	Designer/analyst plays operator	Designer or operator	Multiple operators
Fidelity dimensions	Informational Functional Computational (selective)	Informational Perceptual (i.e., visual displays)	Physical (controls and displays) Anthropometric Perceptual (visual displays)	Communication Informational Physical Computational Perceptual (visual and auditory)
Technical basis	Discrete-event simulation	Manned part-task virtual proto-typing simulation	Manned part-task physical simulator	Networked manned physical simulators
Cost	Low computational software personal computer or workstation	Low computational software graphics software workstation(s)	Moderate graphics software foamcore or fiber-glass mockup	Moderate-to-High networked workstations or physical sim-lators graphics, audio, & computation servers

design process. Table 5.2 provides a comparison of the applicable simulation approaches, their respective strengths/advantages, and cost impacts. As shown in Table 5.2, a staged simulation methodology provides the basis for an effective human–machine integration approach. Each of these simulation stages is discussed here.

Analytic simulation, the first level of simulation, is directed to modeling all "flows" in a manufacturing process with heuristic/rule-based models of human operators working with simulated machine counterparts (Madni, 1988b). From a human–machine integration perspective the purpose of this simulation is to determine operator workload with different function allocation options and levels of automation.

Interactive simulation (i.e., *designer/user-in-the-loop simulation*) is the second level of simulation that employs a real operator (vs. model) working with the simulation. The purpose of this level is to address human–machine integration problems from an operator error and error rates standpoint, uncover deficiencies in the overall concept of operation, and refine the human behavior models within the analytic simulation. This level pertains to the "horizontal prototyping" phase in interactive system design (Madni, 1988b).

Reconfigurable simulator (i.e., *designer/user-in-the-loop prototyping*) is the third level of simulation. The purpose of this level is to evaluate operability of the overall system from an operator perspective using a "virtual prototyping" approach. "Virtual prototyping" uses graphics and audio to replicate the information environment, controls, and displays. It may or may not use simulation to drive the human–system interaction.

Networked simulation (i.e., *multiple users-in-the-loop*) is the level associated with the command and control of a development environment or a factory. From a man–machine integration perspective, this level of simulation analyzes individual operator communication and coordination load and error patterns while operators perform their assigned tasks with different equipment and levels of automation.

Interactive Multimedia Technology

Interactive multimedia technology has opened up a whole new dimension in human–machine relations and human–human collaboration. Specifically, advances in multimedia storage and delivery have expanded the range of options available for teleconferencing, collaborative design, embedded training, and education. Figure 5.4 summarizes a few key capabilities and sample uses of the different components of a multimedia environment. Today, it is widely believed that interactive multimedia technology will realize the much-needed shift in paradigm in systems design and manufacturing. For example, merely incorporating a "live video" window in a design workstation, so that each individual designer can see and hear the others (in a designated window) as opposed to interacting through text messages, can contribute to teamwork and bring about this much-needed cultural change. Similarly, with three-dimensional animation video such as digital video interactive (DVI), computer design-interactive can promote visualization in educational workstations.

Computer-Aided Concurrent Engineering Tools and Their Integration

As designs evolve, the design team needs appropriate tools to support the different design activities. A broad array of analytic, heuristic, empirical, and simulation-based tools are required to achieve the collective objectives of a

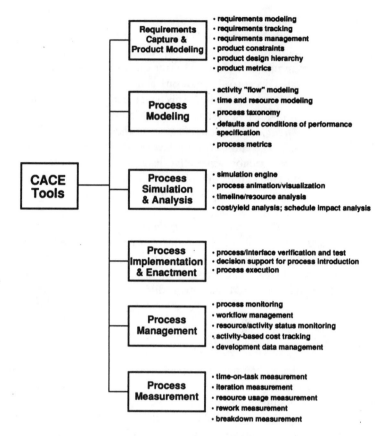

FIG. 5.4.   Interactive multimedia components and usage.

product design within a concurrent engineering rubric. Computer-aided concurrent engineering tools are a set of software tools on graphics workstations that help the collaborative design team visualize and analyze how their design will be built, tested, introduced, and managed on the factory floor. Computer-aided concurrent engineering (CACE) tools are the "productivity multipliers" in product development. Computer-aided concurrent engineering tools serve various purposes and different users. For the tools to be effective they must be delivered on the host environment-compatible platform and language. Figure 5.5 provides an overview of the major elements of computer-aided concurrent engineering.

1. *Requirements capture and product modeling tools* help the user in recording, maintaining, and tracking requirements. The product design hierarchy, constraints, and metrics fall under the purview of this class of tools.

Live video	−promotion of human-machine relations
Stored video	(cultural change)
(compressed in	−embedded help and examples in workstation
education	usage and architecture/design tasks
workstation)	−footage of typical component assemblies, fab line
Computer graphics	−interactive modeling and design of product and
	process (physical abstraction, symbolic metaphor)
	−primary medium of communication
	−display of results (e.g., bar charts, pie charts)
Animation	−dynamic process simulation, alerts in testing, constraint
	violations
Digitized voice	−constraint violation alert in hands-free operator
	environment
Voice	−remote teleconferencing

FIG. 5.5.   Computer-aided concurrent engineering tools.

2. *Process modeling tools* encompass activity "flow" modeling, time and resource modeling, process taxonomy, specification of defaults and performance conditions, and process metrics.

3. *Process simulation and analysis tools* include simulation engines (e.g., discrete event, pattern directed), process animation and visualization, timeline and resource analysis, cost analysis, yield analysis, and schedule impact analysis.

4. *Process implementation and enactment tools* include process/interface verification and test, decision support for process introduction, and process execution software.

5. *Process management tools* encompass process monitoring, workflow management, resource and activity status monitoring, activity-based cost tracking and development data management.

6. *Process measurement tools* include software for recording time-on-task, design iterations, resource utilization, number of resource conflicts, rework measurement, and breakdown measurement.

Despite the existence of numerous tools directed at facilitating the development process, a major impediment to realizing a comprehensive solution is the current state of tool integration. Concurrent engineering software tools may be referred to as integrated merely because they share a common user interface or because a particular vendor offers tools that support more than one phase of the product development life cycle. Some speak of tool integration in terms of the support of shared data storage (Wasserman, 1988), whereas others describe a fully supported development life cycle (Martin, 1990) where all tools interface to a common framework/data base in a distributed environment (Phillips, 1989).

Tools cannot yet be defined as fully integrated, that is, having both control integration and data integration. Most tools/toolsets could be defined as data integrated (or, more appropriately "joined") with an underlying object data base.

The tools may even share this "dictionary," but may be limited in scope, interoperability, or environmental capabilities. These tools may not represent or interact with the multiple views or phases of objects required in coordinated, heterogeneous data storage. Most tools also do not incorporate the external control interfaces and formats that are required for automatic intertool process and data flow.

There are different views on what constitutes tool integration and at what level tool integration should be considered for a specific application or project. Specifically, there are five classes of tool integration:

- Internal integration: integration of tools and data of a single vendor.
- External integration: integration of tools from multiple vendors.
- Environment integration: integration of tools with the operating environment.
- Process integration: integration of tools within development process activities.
- End user integration: integration of tools with end-user groups.

Internal Integration.   Internal integration is the integration of tools and data of a single vendor. Such tools use either a local data dictionary or are "joined" through a central, shared dictionary. The data dictionary is invariably a type of (relational) data base that allows a vendor to offer consistent data for the tools. The resultant product is generally proprietary.

The primary problem with internal integration of single-vendor tools is the limited range of tools offered by the particular vendor. Vendors generally do not offer a full complement of (integrated) tools. Further, the tools tend to be targeted to either personal computers or workstation/mainframe environments. In those rare instances when a single vendor offers a full range of tools, the user may not find the different tools equally useful because of some inherent limitations. Tool users find it unacceptable to be restricted to a single-vendor tool suite due, in part, to these limitations.

The single-vendor aspect also impacts the level of flexibility in a tool set from which a user can benefit. If a vendor allows a user to modify the interfaces/objects of such a nonstandard toolset, it could very well further widen the "compatibility gap" with other (vendor) tools. Even if conversion capabilities are developed to allow a data base/interface to be adapted to a standard format, the ultimate responsibility for adapting the modifications may well be left to the user.

There are also problems inherent in the use of relational data bases as the basis of a software tool data dictionary (Brown, 1989). Relational data bases store data as a collection of logical relationships specified by the user. A data dictionary is a collection of lexicons (lexical units, "words") that provide a common vocabulary by which programs can work together to achieve a task. Relational data bases cannot readily accommodate the flexible complex object/data types (e.g., code segments, design diagrams, user processes) required by computer-aided concurrent engineering technology. They are also generally unable to handle the amount of data that it takes to implement a fine-grained object management system (i.e., objects composed of data items and interrelationships that are more complex than the level afforded by a singular file or character string representation). The problem here is one of data size and performance characteristics. The net result is that (a) users are left with a single, limited view of the data objects, and (b) users find it difficult to share data objects among the tools.

This recognition has spurred many vendors to publish their tool interfaces as a means of promoting greater acceptance from the tool user community (Gibson, 1989; Wasserman, 1988). Whereas this helps users (and other vendors) interface their own tools with the tools of a specific vendor, it fails to address the larger problem of data incompatibility among tools.

Efforts are currently underway to expand the integration of object-oriented data bases (OODB) with tools. An object-oriented data base is the basis of next-generation "repository" products currently under development by both IBM and DEC. A repository is a common shared data base that stores the rules (or relationships) associated with tool data objects. The data may either reside in the repository or be distributed throughout an application network. Object-oriented data bases differ from relational data bases in that they store data maintenance and access rules along with the object data. This technology would provide extended capabilities (e.g., multiple object views, modifiable rules, and object types) and would enhance the distribution/performance aspects of a shared repository. Universal acceptance of object-oriented data base technology is impeded by the fact that the technology is relatively new (no single data model) and by the prior commitment to relational data bases of many vendors, users, and standards organizations.

External Integration.   External integration pertains to the integration of tools from one vendor or with those of another vendor. This integration is usually in the form of "access control" and/or "data control." With access control, the tools of one vendor can be invoked by tools from other vendors. The tool invoked returns appropriate messages/codes to the invoking process/tool. With data control, external tools are allowed to indirectly manipulate the data contained in the internal dictionary of a tool.

Due to the development of proprietary interfaces, most vendors cannot readily interface with other vendor data bases. Not only might the format of the data be different, but the contents of the dictionaries may be inconsistent. The weaknesses of relational data bases are also exacerbated when objects are transformed. Because data rules are imposed by the tools (as opposed to the data base), the consistency of objects (the "view") may well be distorted as semantic content is lost/misinterpreted when moving data between tools.

Another problem with external integration directly underlies the importance of the selection and acceptance of standards. The integration of the tool/data interfaces from two different vendors is a costly proposition that includes both the indirect support and maintenance associated with the tools and interfaces of both vendors. Ultimately, given the current state of standardization efforts, the end users themselves may have to act as systems integrators and write the code necessary to effectively integrate a tool from different vendors. This assumes that tool interfaces are well documented and that the user can afford the integration of time/cost.

Work is currently underway by IBM, DEC, and others to define a central data repository that would take steps to solve the data incompatibility issue.

Environment Integration.   Basically, tools are tied to their environment through the host operating system or hardware platform. The tools may use specific features of the operating system or support tool set that are unavailable on other systems (e.g., windowing system variants, Unix system variants). This makes it harder (and in some cases impossible) for these tools to be rehosted to other environments.

Problems also arise with scalability when considering a move from one environment to another. Some tools might experience performance degradation or bump up against environment constraints (e.g., memory requirements, CPU characteristics, data storage facilities). Tools developed for a single-user PC environment may be unable to support multiuser projects because of concurrency or project/data size requirements. For example, it may be impossible to run multiple instances of the tool or the tool may be unable to handle the increased data access requests.

Another environment-related issue is that of tool integration with a configuration management system. A configuration management system is a framework in which a set of objects (files, programs) can be logically grouped into a working unit. No viable tool offering (regardless of the scope of the tool set) can overlook the importance of maintaining an ordered version control history for data objects. Some tools integrate a simple versioning scheme for file objects, but the concept must be extended to include simple versioning scheme for file objects, but the concept must be extended to include all data types as object granularity becomes finer than the file level and as source code becomes more a "derived object" than merely a central entity. Extension of tools to include configuration support must be given careful consideration so as not to trigger scalability problems (particularly data storage and performance limitations).

One possible solution to the environment integration issue is the emergence of a standard for the Unix operating system. Unix has been showing promise as a "crossover" operating system catering to the needs of both the technical and commercial markets (Cortese, 1990; Cureton, 1988). In this respect, tool vendors would be relieved of the burden of maintaining separate product lines for multiple operating systems by targeting the Unix system. This would also alleviate many of the rehosting issues facing the vendors as product porting would become solely a hardware (Unix system platform) issue. Environment integration and tool/data base distribution would be greatly enhanced through the accessibility of existing network support functions (e.g., network protocols).

Process Integration.   In anticipation of smoothing life-cycle process transients, software tools are being developed to help combine the various phases of the entire life cycle (e.g., project management, analysis and design, configuration management).

The integrated project support environment (IPSE) offerings allow for tool data, control, and presentation integration. Data integration refers to the coordination of access to the underlying tool data base(s). Control integration refers to the coordination of access to the tools themselves. Presentation integration refers to the coordination of the user interface. The basis of data integration is the repository. The basis of control integration is the software "backplane" or executive that provides the requisite interfaces to the different tools. Integrated project support environments allow for mixed operating system support and, in some cases, for mixed platform support.

A key issue in process integration is that of tool flexibility. Users demand that tools be easily modifiable and customizable to their particular needs. Most tools remain closed to such customization. Those that claim an "open" interface generally allow modification of merely the presentation characteristics. For tools to be fully integrated, users must be able to modify the characteristic behavior of the tool (e.g., design rules, object rules, object types, process), not just the user interface (Forte, 1989). Users should not have to abide by a strictly imposed process for software development (e.g., waterfall model, spiral model) if they are best served by some internally developed or hybrid process.

Another key problem is that most tools support only very specific design methodologies. Also, the methodology supported by the tools is usually strictly enforced. Consequently, the users have to select and learn a new methodology or choose from a limited set of tools that support their current methodology. In the interest of accommodating in-place methods and procedures, tools need to be able to expand beyond the traditional "bottom-up" or "top-down" design approaches and allow configuration modifications to support alternative ("middle out") methods.

Another term used in the tool integration arena is *framework*. Frameworks are essentially tool backplanes with a tool management executive. The executive handles the overhead involved with coordination of the tool suite (e.g., user presentation, tool registration and instantiation, error reporting). At the bottom end of the framework is a common data interface/repository, messaging system, and operating system services manager. These interfaces unburden the tool modules of the particulars of the host environment and allow for a broader, more interchangeable product set.

End User Integration. The concept of integration with the end user ranges from something as simple as maintaining a consistent user interface to something as complex as providing support for an expert system interface. In particular, a standardization (e.g., Open Software Foundation/Motif) could help enhance user acceptance of tools.

One of the basic concerns of tool integration environments in general is the issue of a consistent user interface/presentation (Forte, 1989; Phillips, 1989). The learning curve associated with adopting a new tool is steep and requires a significant investment on the part of an organization that goes well beyond the cost of the tools. It involves the time and cost of comprehensive training and support (both from the vendor and from the users). Standardized user interfaces and tool functionality could help keep these costs under control as well as offer greater tool/choice flexibility to the users.

In addition to the underlying system, user interface consistency is also becoming a key issue to tool vendors. The user interface contributes significantly to the acceptance and learning time for a product. It would seem that the user interface is the easiest of the integration areas on which to standardize; yet vendors must use caution not to promote a "quick and dirty" solution. If the vendors do not carefully analyze the requirements of the user interface and miss taking into account the usage context of the tool, they may end up with an interface standard that promotes consistency at the expense of ease of use.

Technology Introduction Strategy

As with any new technology, the introduction of concurrent engineering methods and, more specifically, computer-aided concurrent engineering and computer-supported collaborative work can be expected to face some degree of resistance from users despite the potential benefits.

Gould and Lewis (1985) recommended that "early in the development process, intended users should actually use simulations and prototypes to carry out real work, and their performance and reactions should be observed, recorded, and analyzed" (p. 300). This is all the more significant for computer-supported collaborative work because of the added task complexity. When the development cycle involves individual users performing single tasks (e.g., using a spreadsheet application), design errors emerge relatively quickly because the interactions are limited to one person and one system. However, when the system supports the cooperative work of multiple users, a higher level of complexity is involved. Interactions among multiple users create a set of interdependencies not found in single-user systems. As a result, design errors emerge more slowly and are more difficult to pinpoint. Additionally, there is a greater opportunity for unintended

TABLE 5.3
Strategies for Large-Scale Implementation

1. Demonstrations of the technology for potential users.
2. Group consensus on the goals for the technology.
3. User participation in development of an implementation plan.
4. Implementation on a trial basis in one area of the organization.
5. Evaluation of the implementation trial.
6. Revision of the implementation plan (if needed).
7. Monitoring and periodic review of the implementation effort.

efforts, some of which may not appear for a long time! This performance-based analysis approach is key to facilitating technology transition.

Also, human factors will play a significant role in ensuring the acceptance of concurrent engineering and computer-supported collaborative work. For small-scale implementation, implementation guidelines and support training courses should be instituted. For large-scale implementation, on-site consulting may prove to be much more cost-effective. Geirland (1986) recommended including seven strategies in large-scale implementation efforts (Table 5.3).

Antonelli (1988) suggested that design decisions are more sound when based on input from real users. In this regard, the identification of cooperative work modes (P. Johnson-Lenz & T. Johnson-Lenz, 1982) is useful for developing computer-supported collaborative work tools. One such taxonomy is presented in Table 5.4.

In light of the foregoing considerations, we have identified the key elements of a successful approach to technology introduction within product development and manufacturing environments (Table 5.5). This approach is grounded in the key concepts summarized in Table 5.5. Each of these concepts is discussed in the following paragraphs.

TABLE 5.4
Taxonomy of Cooperative Work Styles in Teleconferencing

1. Individual work versus group interaction.
2. Anonymity versus signed responses.
3. Feedback of group results versus no feedback.
4. Aggregated versus no voting.
5. Voting versus no voting.
6. Numerical processing versus symbolic processing
7. Filtered information (selection mechanisms to access only selected items) versus unfiltered information.
8. Synchronous versus asynchronous interaction.
9. Sequenced versus free or unstructured interaction.
10. One-time access to information versus continuous access.
11. Patterns of communications: one-to-one, one-to-many, many-to-many, many-to-one.

TABLE 5.5
Transition Strategies

Start small, then expand
Staged series of demonstrations
Indoctrination of end users
Storyboarding
Horizontal prototyping
Vertical prototyping
Exploitation of inplace procedures and tools
End users involvement in all phases

Start Small, Then Expand

Taking on the total manufacturing environment as the target environment is much too big a task. One strategy to overcome this problem is to identify high payoff targets of opportunity for the insertion of concurrent engineering methods and tools. This approach not only makes the problem tractable, but also gives some evidence of where to set our sights next. Successful insertion of concurrent engineering within certain key processes can potentially facilitate the generalization of lessons learned to other functions such as assembly and integration.

Staged Series of Demonstrations

A key strategy to communicate objectives and forthcoming changes is through a series of demonstrations. Typical demonstrations might include product development; process model development and documentation; process simulation and visualization; process enactment in a limited setting and process management in computer-supported collaborative work; tool usage; tool interoperability; and integrated factory subprocess simulations.

Indoctrination of End Users

Given the heterogeneous nature of the design team, and the change of culture required for successful introduction of concurrent engineering, it is imperative that the end users be provided with the "big picture" along with the specific objectives and their respective roles in the design process. Without a common understanding of end objectives, and preparation for change, the user resistance can be formidable. At the level of individual tools, it is equally important to show end users how the tool fits in the overall "new" process.

Storyboarding

The objective of the storyboarding phase is to develop a set of sequential, static interface screen layouts corresponding to the preliminary concept of operation of a tool, a device, or a subsystem (Madni, 1988b). These series of screens with appropriate textual and pictorial annotations serve as the basis for communicating the functionability of the tool to potential users. Specifically, the screens serve to communicate to the user what the tool/system does versus what the user is expected to do. Storyboarding provides the first indication of the level to which the tool/system can be expected to aid, train, or off-load the user. In addition, the initial screen compositions become a point of departure for soliciting user comments on improving the utility and composition of the screen layouts—the user identifies missing information and/or all information elements that are best presented as exceptions. At this stage, users can indicate the specific types of help and/or software alerts they prefer as they "walk through" the tool/system storyboards. In sum, storyboarding serves as an effective means for specifying interactive software operation. Prototyping efforts can start once all pertinent user comments are incorporated into the storyboards.

Horizontal Prototyping

This particular prototyping strategy pertains to the high-fidelity replication of the user interface of the final system with simulated functionality and durations (Madni, 1988b). The purpose of horizontal prototyping is to provide a vehicle for identifying shortcomings and errors in user–system interactions. Insofar as concurrent engineering tools are concerned, these deficiencies are in the form of missing or extraneous information, inordinate time delays in user–system interactions, suboptimal windowing and screen layouts, and so on. Horizontal prototyping greatly enhances the tool's overall usage concept and supports early demonstrations of the evolving functionality of the tool.

Vertical Prototyping

Vertical prototyping is the high-fidelity implementation of selected functions for user examination and feedback. The purpose of vertical prototyping is to define and implement in detail those functions that are important to overall system operations and for which user inputs are critical to improving system implementation (e.g., assumptions, algorithms). Horizontal and vertical prototyping can often be done concurrently with the results presented in the same demonstration (Madni, 1988b).

Exploitation of In-Place Procedures and Tools

One of the key concerns in the introduction of concurrent engineering methods, practices, and tools is that one does not disrupt ongoing activities and working in-place procedures. To this end, an analysis of procedures and tools in use within the manufacturing environment should be undertaken with a view to tailoring the introduction of concurrent engineering, to the extent possible, to either subsume these procedures and tools or be compatible with them.

End User Involvement in All Phases

A key concern in the introduction of new technology is user acceptance. To bias the odds in this area, end users should be involved as key contributors in all phases of design. This strategy will not only increase user acceptance but, in fact, highly effective solutions may very well come from end users who have an informal data base of lessons learned.

CONCLUSIONS

Concurrent engineering has emerged as a new way of doing business in the design and manufacture of new products. Although concurrent engineering has been accepted in concept, its full impact can only be felt when all the enabling technologies and tools are in place. This chapter discusses collaborative design environments, process modeling and simulation, human–machine integration, interactive multimedia technology, and computer-aided concurrent engineering tools and their integration—five key technological components that must be implemented prior to successfully introducing concurrent engineering approaches, practices, and procedures.

Collaborative design environments are key to supporting teamwork but some technical hurdles (e.g., the problem of sharing design objects) have to be overcome. Process modeling and simulation potentially provides members of the design team with the ability to "look down the line" while still in the conceptual design phase. However, modeling, integrating, and displaying all the different "flows" in a manufacturing enterprise without overwhelming the users continue to be a major change. A systematic methodology for human–machine integration is key to successful human–machine performance. The suggested simulation-based approach to human–machine integration is both methodical and cost-effective. Interaction multimedia technology holds great promise as a motivator and a precursor to the required cultural change. The specification of a computer-aided concurrent engineering tool kit that spans the life cycle of the product will not only reduce designer time-on-task, but also provide an audit trail of design decisions and "lessons learned." Although tool integration continues to be a major challenge, the emergence of standards, should make this problem more tractable.

ACKNOWLEDGMENTS

The research reported in this chapter was supported in part by a contract from the Defense Advanced Research Projects Agency (DARPA), administered by the Office of Naval Research (ONR), to the UCLA Center for the Study of Evaluation/Center for Technology Assessment. However, the opinions expressed do not necessarily reflect the positions of DARPA or ONR, and no official endorsement by either organization should be inferred.

REFERENCES

Antonelli, D. C. (1988). Research and product usability. *Human Factors Society Bulletin, 31*(11), 2–4.

Brown, A. W. (1989). *Database support for software engineering.* New York: Wiley.

Cortese, A. (1990, January). CASE standard comes overseas for Unix arena. *Computerworld,* Vol. 24, No. 2, pp. 23–24.

Cureton, B. (March, 1988). The future of Unix in the CASE Renaissance. *IEEE Software,* Vol. 5, No. 2, pp. 18–22.

Estrin, G., Fenchel, R., Razouk, R., & Vernon, M. (1986, February). SARA (System ARchitects Apprentice): Modeling, analysis, and simulation support for design of concurrent systems. *IEEE Transactions on Software Engineering, SE-12,* 293–311.

Forte, G. (1989). In search of the integrated CASE environment. *CIA/S/E Outlook, 89,* (2).

Geirland, J. (1986). Macroergonomics: The new wave. In *Office systems ergonomics report,* Vol. 5, (pp. 3–11). Santa Monica, CA: The Koffler Group.

Gibson, S. (1989, March). Some win, some lose when repository debuts. *Computerworld,* Vol. 23, No. 11, p. 141.

Gould, J. D., & Lewis, C. (1985). Designing for usability: Key principles and what designers think. *Human Aspects of Computing, 28,* 300–311.

Harel, D., Lachover, H., Naamad, A., Pnueli, A., Politi, M., Sherman, R., & Shtul-Trauring, A. (1988, April). STATEMATE: A working environment for the development of complex reactive systems. IEEE Transactions on Software Engineering, Vol. 16, No. 4, pp. 403–415.

Johnson-Lenz, P., & Johnson-Lenz, T. (1982). Groupware: The process and impacts of design choices. In E. G. Kerr & S. R. Hiltz (Eds.), *Computer-mediated communication systems,* (pp. 45–55). New York: Academic Press.

Madni, A. M. (1988a). HUMANE: A designer's assistant for modeling and evaluating function allocation options. *Proceedings of Ergonomics of Advanced Manufacturing and Automated Systems Conference.*

Madni, A. M. (1988b). The role of human factors in expert systems design and acceptance. *Human Factors Journal, 30*(4), 395–414.

Madni, A. M., Balcerak, R., Estrin, G., Freedy, A., & Melkanoff, M. A. (1990, February). *Computer-aided concurrent engineering (CACE) of infrared focal plane arrays (IRFPAs): Emerging directions and future prospects.* Paper presented at the Second National Symposium on Concurrent Engineering, Morgantown, WV.

Martin, J. (1990). Integrated CASE tools a must for high-speed development, *PC Week,* Vol. 6, No. 3, p. 78.

Mujica, S., Berson, S., Estrin, G., Eterovic, Y., Leung, P., & Wu, E. H. (1990). *Architecture for sharing design objects in the UCLA SARA collaborative design environment.* Los Angeles: Computer Science Department, University of California.

Phillips, B. (1989). A CASE for working together. *ESD: The Electronic System Design Magazine,* Vol. 19, No. 12, pp. 55–57.

Wasserman, P. I. (1988). Integration and standardization drive CASE advancements. *Computer Design,* Vol. 27, No. 22, p. 86.

Winner, R. I., Pennell, J. P., Bertrand, H. E., & Slusarczuk, M.M.G. (1988). The role of concurrent engineering in weapons system acquisition (Institute for Defense Analyses Internal Technical Rep. No. R-338, Task T-B5-602). Alexandria, VA: Institute for Defense Analyses.

6 Assessment of Software Engineering

Kathleen M. Swigger
University of North Texas

In 1967, a NATO study group was formed to discuss the "software crises." At the end of 1 year, the group concluded that building software is similar to other engineering tasks and software development should be viewed as an engineeringlike activity. Thus, the phrase "software engineering" was born, along with the belief that programing was simply the application of certain scientific and engineering principles. As a result, texts were written and metrics established for the purpose of identifying the scientific principles of software engineering (Gelperin & Hetzel, 1988). The fact that programs still contain bugs, are delivered late, and are overbudget should tell us that many of the basic scientific principles of programing remain undiscovered. Yet the goal remains that software engineering is a discipline whose aim is the production of quality software, software that is delivered on time, within budget, and that satisfies its requirements. As a result, most software engineers interpret technology assessment as the set of procedures that are used to determine "how well" the software meets its *initial* specifications. On the other hand, most software users maintain that technology assessment (i.e., software assessment) should be a measure of how well the software meets the *user's* needs.

In order to meet each of these goals, the scope of software engineering has become extremely broad, encompassing every phase of the software life cycle, from requirements to decommissioning. It also includes different aspects of human knowledge such as economics, social science, and psychology. To this end, a variety of techniques have been developed for performing and evaluating various software production tasks, from requirements and specifications to maintenance. In addition to measuring the quality of software, there are numerous studies that compare different techniques and methodologies used to write, comprehend, and debug software. As a result, the relatively new challenge for software engineers is to develop assessment techniques that work and possibly reflect the more human aspects of software development, those that acknowledge the importance of both the programer and the user.

The purpose of this chapter is to review some of the major techniques that have been developed for assessing software production tasks and to show how these techniques might be applied to technology assessment in other areas. To this end, this chapter has three major themes. The first is analysis and comparison. A variety of techniques are described for software evaluation and the evaluation of software engineering. Software evaluation is defined as the assessment of a specific piece of code or program produced by individuals and/or a team of programers. On the other hand, software engineering refers to the practices, techniques, and procedures used to produce quality software. Because of the variety of present-day software engineering techniques, it is important to select an appropriate one for the task at hand. A second theme is that the results of experiments in software evaluation and software engineering constitute a powerful tool for determining which techniques are useful for a given situation. These same techniques and tools might be used by other researchers in related fields of technology. The third theme is that the future of software production will necessitate the development of a new definition of software engineering that recognizes the human aspects of both software development and its evaluation.

In order to develop these themes, the chapter focuses on different perspectives of software engineering. The next section introduces the software evaluation perspective. The wide scope of different measures used to evaluate software is highlighted, as are the problems of conducting software evaluation tests. The third section compares and contrasts different software engineering methodologies. Experiments related to the different methodologies are also reported. The fourth section includes a discussion of current software engineering topics, such as computer-aided software engineering (CASE) products and computer-supported cooperative work environments (CSCW), and shows how each of these topics reflects an understanding of human problem solving. These topics were selected because of their relevance both to current software engineering practices and to the theme of this book. The chapter concludes with a brief summary and discussion of future research.

SOFTWARE EVALUATION

Definition and Characteristics

Since the 1970s the distinction between software and software engineering has become blurred. The distinction was very clear in the "good old days." Software was a single piece of code that was compiled and executed. On the other hand, software engineering referred to the set of systematic procedures that were applied and used when developing a collection of programs (Schach, 1990). Now, however, the creation of an isolated piece of code is extremely rare. Most software is written by teams of programers who access and control several other programs or hardware. However, in order to distinguish among the different software evaluation techniques, the rest of the chapter uses the term *software* to denote the end result of a process (i.e., a product), whereas software engineering refers to the process of developing the software. Thus, software evaluation involves the process of ensuring that the actual product itself is correct (i.e., does not contain a fault or bug and meets its specification), whereas evaluation of software engineering involves testing whether the procedures used to produce the software were successful in creating an acceptable product.

Software quality measurement is a young discipline. Because of its relative youth, there are conflicting opinions as to what and how specific software characteristics should be measured. For example, several authors have argued that software evaluation means testing a program until it no longer contains any bugs. Unfortunately, as Dijkstra (1976) pointed out, "Program testing can be a very effective way to show the presence of bugs, but it is hopelessly inadequate for showing their absence" (p. 859). Software evaluation, therefore, implies the testing of a product to determine if it is correct (program verification) and exhibits certain behavioral properties (program validation).

The primary goal of any testing procedure is to determine whether the product functions correctly (i.e., does not contain a bug). Additional software characteristics include utility, reliability, robustness, performance, and correctness (Gelperin & Hetzel, 1988). *Utility* refers to the extent to which a program meets the user's needs, given that the product is used according to its specifications. The utility of the product is determined by verifying that the program produces correct outputs when subjected to inputs that are valid in terms of its specifications.

Reliability is a measure of the frequency and criticality of product failure, where failure is an unacceptable effect or behavior occurring under permissible operating conditions. Software reliability is calculated by adding the mean time between failures to the mean time for repairing the failures.

A product's *robustness* is a function of the range of operating conditions, the possibility of unacceptable effects on valid input, and the acceptability of effects when the product is given invalid input. A fourth characteristic, *performance,* is defined as the extent to which the product meets its constraints with regard to response time, execution times, or space requirements.

The last characteristic, *correctness,* refers to a mathematical procedure used to prove that a product satisfies its output specifications, when operated under permitted conditions (Goodenough, 1979). In other words, the product is correct if the output satisfies the output specifications, given that the product has all necessary resources.

```
UTILITY
     *Usability tests

RELIABILITY
     *Walkthroughs
     *Code inspection
     *Number of bugs + downtime
        between failures

ROBUSTNESS
     *Test case selection
     *Functional testing

PERFORMANCE
     *Execution time
     *Average response time
     *Memory constraints
     *Hardware constraints
     *Portability

CORRECTNESS
     *Proofs of correctness
```

FIG. 6.1. Measurement techniques for program characteristics.

Having identified the important software characteristics, it becomes necessary to construct appropriate tests that measure each of the different characteristics (Fig. 6.1). Unfortunately, different types of software emphasize different software characteristics. For example, programs using artificial intelligence (AI) techniques emphasize performance criteria rather than accuracy or optimality (Brazile & Swigger, 1990). Indeed, speed of execution is often the major reason for selecting an AI technique over other conventional methods. In an AI program, specialized knowledge rather than exhaustive search is used to derive quick solutions to specific problems. In contrast, operations research techniques are designed to produce optimal results and, as such, require the program to use an exhaustive search to find the "best" solution. Because of space limitations, it is impossible to list all the testing procedures that are used in combination with the various types of software. The remainder of the section, therefore, describes only a few testing techniques used by program developers. The reader is encouraged to investigate other sources for additional information (e.g., Gelperin & Hetzel, 1988; Perry, 1983).

Walkthroughs and Code Inspections

During the design and specification phases of the software development life cycle, it is vital to verify that the specifications are correct. The best way to do this is by means of a walkthrough or code inspection. A *walkthrough* is a software review performed by a team of software professionals with a broad range of skills (Shneiderman, 1980). The team is usually comprised of four to six members who are charged with the task of offering an unbiased report of the software under construction. Thus, the lead programer "walks" the other members through the program. The members interrupt either with prepared comments or with questions triggered by the presentation. In this manner, the software is examined for faults or irregularities.

A *code inspection* is another form of software review. The team reviews the program specifications against a prepared checklist that includes items such as the following: Have the hardware resources required been specified? Have the acceptance criteria been specified? A code inspection is a more formalized review process than a walkthrough and usually involves five basic steps (Fagan, 1976). First, an overview of the design is given by the lead programer. Second, the code inspection team prepares comments individually for the inspection. Third, the group inspects the code that involves addressing each of the individual comments and ensuring that every piece of logic is covered at least once, and every branch is taken at least once. The fourth step, called rework, involves resolving all faults and problems. The final stage involves a follow-up in which the inspection team ensures that all questions have been satisfactorily resolved.

Selection and Use of Test Cases

In order to verify that the program functions correctly, a group of test cases is constructed to either test to specifications (also called black-box, data-driven, functional testing) or test to code (also called glass-box, logic-driven, or path-oriented testing). Using the former technique, the programer constructs a series of test cases that correspond to the software's specifications (Perry, 1983). In contrast, the test to code technique requires the programer to construct test cases that consider only the code itself. Regardless of which technique is selected, a "complete" testing program requires the construction of literally millions of test items. Therefore, the "art" of test case construction is to find a small, manageable set of test cases that maximize the chances of detecting a fault while minimizing the chances of having the same fault detected by more than one test case (Myers, 1978). In addition, the program developer may also have to formulate test cases that meet specific user-supplied criteria in the form of benchmarks. Over the years, benchmarks have been developed for the purpose of comparing one computer architecture (or software) to another objectively. Thus, whenever a new computer system (or software) is proposed, the end user specifies a series of benchmarks that are designed to show that a proposed system can meet stated its performance criteria. In order to satisfy the benchmarks, the program developer or vendor must create a series of test cases showing that the software can meet or satisfy the proposed benchmarks.

Testing to Specifications

Equivalence Testing and Boundary Value Analysis.   To determine if the product runs correctly, the program designer constructs a set of test cases such that any single member of the class is representative of (or equivalent to) all other members of the class (Schach, 1990). For example, if a program is written to handle a range of numbers between 1 and 35, then the programmer defines three different equivalence classes:

Equivalence class 1: numbers less than 1.

Equivalence class 2: numbers between 1 and 35.

Equivalence class 3: numbers more than 35

Testing a program using the equivalence class technique requires constructing one test case from each equivalence class. The test case from Equivalence class 2 would produce the correct answer, whereas the test cases from the other two classes would produce error messages.

Experience has shown that, when a test case is selected from either side of the boundary of an equivalence class, there is a high probability of locating a fault. Testing the previous example using this technique (i.e., boundary value analysis) would produce seven different test cases:

Test case 1: 0 (number adjacent to boundary condition)

Test case 2: 1 (boundary value)

Test case 3: 2 (number adjacent to boundary condition)

Test case 4: 15 (member of equivalence class 2)

Test case 5: 35 (boundary value)

Test case 6: 36 (number adjacent to boundary condition)

Test case 7: 37 (number adjacent to boundary condition)

Equivalence class testing combined with boundary value analysis is an effective technique for locating faults and requires a relatively small set of test data. Research has shown that, when used together, these two methods constitute an extremely powerful evaluation tool (Basili & Selby, 1987).

Functional Testing.   An alternative to the previous technique is to construct test cases based on a program's functionality (Howden, 1987). After defining the program's functions, test data are then constructed such that there is at least one test case for every function in each module in the program. If the modules are designed in a hierarchical fashion, then functional testing proceeds in a bottom-up manner. In practice, however, modules and subroutines are highly interconnected and require complex functional testing techniques; for details, see Howden (1987).

Testing to Code

Structural Testing: Statement, Branch, and Path Coverage. The simplest form of code testing is to examine each individual statement (i.e., *statement coverage*), and construct a test case that ensures that every statement is correctly executed (Schach, 1990). Usually an automated tool is required to record which statements have been executed over a series of tests. *Branch coverage* is another type of functional testing and involves running a series of tests to ensure that all branches in the program are executed at least once. The most powerful form of structural testing is *path coverage,* which requires testing all paths through the program. Unfortunately, if the product contains many loops, the number of paths through a program can be computationally quite large. Thus, the programer learns to reduce the number of paths by restricting test cases to only linear code sequences (Woodward, Hedley, & Hennell, 1980), or to code sequences that lie between the declaration of a variable and its use (Rapps & Weyuker, 1985).

Complexity

Software complexity measurement is an attempt to merge theories from cognitive psychology with theories from computer science. Similar to psychological measures, software complexity refers to a program's characteristics that make it difficult for a human to understand. The underlying assumption behind these metrics is that a product's complexity is a good predictor for product reliability, performance, and flexibility (Shneiderman, 1980). Thus, if the complexity of a program is measured and found to be extremely high, then the module should be rewritten because it will be cheaper and faster to start all over than to try and debug and maintain the existing code.

The simpliest and most frequently used measure of complexity is *lines of code.* Although this type of measure has proven ineffective for determining programer productivity, it can be a useful predictor of the number of faults in a program (Basili & Hutchens, 1983; Takahashi & Kamayachi, 1985).

Other, more accurate predictors of product complexity and fault rates look at either the number of decision points in a program or the number of operators and operands. For example, McCabe (1976) developed a measure of complexity based on the number of branches in a program. He also argued that the complexity of a program consisting of N modules is the sum of the complexity of the individual modules. McCabe's metric can be computed almost as easily as lines of code and has been shown to be a good predictor for the number of faults in a program (Schach, 1990).

Halstead's software science metric (Halstead, 1977) has also been used to measure complexity and fault rates. Halstead's method is based on the ability to count, for any program, the number of unique operators (such as IF, =, DO, PRINT) and the number of unique operands (such as variables or constants). The other two basic elements are the total number of operators and the total number of operands. From those counts, Halstead derived functions for predicting properties such as program length, program volume, and program level. Subsequently, he used those results with theories and assertions relative to cognitive psychology, such as Miller's (1956) study on the brain's capacity for processing information, and derived equations that predict the mental effort and time required to write different programs. For example, he hypothesized that the optimal volume for a program should not exceed seven (plus or minus two) and performed several studies that seemed to substantiate this belief. He also speculated about the relationship between program metrics and text analysis.

Although the idea of measuring a program's complexity is appealing, the exact nature of its use remains in question. The development of a theory of programing based on the most primitive components of programs—operators and operands—unquestionably is appealing. However, the ability to provide measures that accurately reflect and predict the mental processes involved in programing has not been fully documented.

Correctness

Recently, the idea of correctness proofs has become a major topic in computer science (Dijkstra, 1990). As the name implies, a correctness proof is a formal mathematical logic indicating that the program meets its specification. In an an attempt to provide a mathematical framework for computer programing, several researchers have developed special verification techniques that prove the correctness of a program.

The major difference between testing and correctness proofs is that testing is performed by executing a program, whereas a correctness proof is a mathematical verification that the product is correct; the proof is never executed using a computer. A program is said to be correct if its output satisfies its input specifications. In order to prove a program correct, the programer uses a mathematical notation, similar to that used by theorem provers, to show that each step of the program proceeds in a logical sequence and conforms to the stated problem definition. This, of course, does not necessarily mean the program is acceptable to the user. It only means the program satisfies its specifications.

Space requirements prohibit an elaborate explanation of correctness proofs. A brief example of a code fragment along with its corresponding flow chart and proof is provided for the interested reader (Fig. 6.2). Additional details on how to perform correctness proofs can be found in Manna (1974) and Dijkstra (1976).

It has been shown that correctness proofs by themselves are insufficient to verify programs. It has been demonstrated that a program that has been proven correct may still contain errors (Goodenough & Gerhart, 1975; Leavenworth, 1970). The studies indicate that the combined use of test cases and correctness proofs is the only way to ensure that a program contains no faults. Thus, correct-

CORRECTNESS

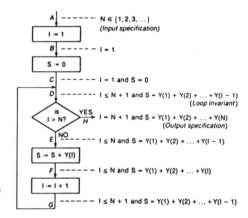

FIG. 6.2.   Example of a proof of correctness.

ness proving must be viewed as belonging to the larger set of techniques that can be used to check that a product is correct.

Implications of Software Testing for Technology Assessment

A number of studies have been performed comparing different strategies for testing software. For example, Myers (1978) compared specification-testing techniques with different code-testing techniques and structured walkthroughs. Similarly, Basili and Selby (1987) compared specification-testing, code-testing, and code inspection techniques. Both studies found that different testing techniques were equally effective in finding the faults in programs, with each technique having its own unique strengths and weaknesses. Although no one technique was found to be superior, they were all found to be better than using no technique.

The obvious implication of such studies is that educational software (as well as other types of application software) is *software* and, as such, requires testing similar to that described earlier. The application of software evaluation techniques to the evaluation of educational software should be performed at each stage of the software development process. It is interesting to note, for example, that educational researchers routinely report measures of validity and reliability for their software as it relates to enhancing IQ or skill acquisition. Yet, these same researchers rarely report measures of program reliability or validity for the code in the educational software. The absence of such measures seems to indicate that authors of educational software are either ignorant of software testing procedures or unable to perform such tests. We hope this is not the case and that program reliability and validity for educational software will be reported in the near future.

Another implication of the software evaluation process is that software testing techniques, especially those used for the construction of test cases, can be applied to other areas of technology assessment. For example, in order to demonstrate that software is appropriate for different types of populations, it is sufficient to show that members from different equivalence classes can perform equally well.

Although software testing may ensure that the product contains no faults, none of the aforementioned techniques ensures that the user will like the product. Software evaluation measures simply test for errors and violations of specifications. If the specifications are poorly defined, then the software may be unacceptable to the user. As a result, the software evaluation process should consider the context in which the product is used and produced.

EVALUATION OF SOFTWARE ENGINEERING

The Life-Cycle Models

As previously stated, the idea that a product exists as a single piece of code is no longer valid. A program may be required to run in parallel, on different machines, under different operating systems, and accessing multiple databases. As a result, special methodologies, models, and procedures are used to systematize the production of large programs. The broad term assigned to these approaches is the "life-cycle model" because it describes procedures for carrying out the various functions of software development. Once a model has been selected, milestones and an overall plan for product development are established.

The Waterfall Model. Two common software engineering models are the waterfall model and the rapid prototype model. The models are very similar to each other and vary only in one area.

As first proposed in 1970, the waterfall model (Royce, 1970) describes the conventional approach to software development. The version, as it appears in Fig. 6.3, suggests that the software development cycle consists of six separate

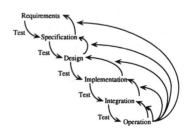

FIG. 6.3.   Waterfall model.

phases: requirements, specifications, design, implementation, integration, and operations. Following each phase is a period of testing and verification that, if unsuccessful, forces the developer to reevaluate previous specifications and design. The waterfall model with its feedback loops and iterative design process allows for revisions of the design, and even the specifications, at every stage of the process. It should be noted that testing is not a separate phase of the process, but occurs continuously throughout the life cycle of the product.

Although communication with the client occurs at each phase of the life cycle, a list of specifications does not always tell the user how the finished product will look or feel. Thus, the waterfall model, depending on how it is implemented, can sometimes lead to software that is unacceptable to the user. As a result, the rapid prototyping model was developed to solve this problem.

The Rapid Prototype Model.   A prototype is a working model that is functionally equivalent to a subset of the product. The first step in the rapid prototyping life cycle is to specify a product's functionality and then build a program that matches those specifications. Once the client is satisfied, the software development process continues, as shown in Fig. 6.4 (Schach, 1990). The two most important items to remember when using the prototype model are that (a) the prototype is built for change, and (b) the prototype is built as input to the specification stage. Thus, the prototype is simply a minor detour from the normal path of software development (Rothenberg, 1990).

The use of rapid prototyping as a way of minimizing risk is the idea behind the prototype model. Unfortunately, rapid prototypes are often accepted as the end product, or as a substitute for written specifications. Another potential problem is that prototypes may not adequately assess hardware needs for large-scale software products. There are substantial differences between large- and small-scale software, and a prototype cannot adequately assess the type of hardware needed for large-scale tasks.

Yet rapid prototyping combined with the waterfall model can produce an acceptable life-cycle model for developing software. Prototypes are extremely useful for demonstrating how the interface will look to the end user. On the other

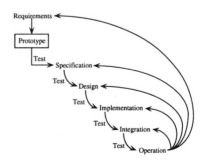

FIG. 6.4. Rapid   prototype model.

hand, the waterfall model provides a systematic set of procedures for designing, implementing, and integrating large-scale products.

The Implications of the Life Cycle to Technology Assessment. As previously stated, testing is an inherent component of the entire product life-cycle and requires careful validation at every stage of development. Following each phase of product development, specific tests are performed and examined. For example, a structured walkthrough is staged during the specification stage, whereas module testing and test case selection occurs during the implementation phase (Fig. 6.5). The idea that different tests are used at different times during product development should be applied to other areas of technology assessment. It is not uncommon, for example, to find a research team comparing students' performance using different media. These studies consist of a single test designed to determine whether the technology meets its requirements, is integrated correctly, and outperforms all other treatments. A better approach to software evaluation is to design multiple tests in parallel with every stage of product development.

The Design Phase

Rather than describe each phase of the software life cycle, together with its appropriate testing procedures, this section focuses on the *design phase* of software development. Just as an an outline serves as a catalyst for written works, a design technique drives effective software development. Thus, program design is

TESTING DURING REQUIREMENTS PHASE
 *Protype development

TESTING DURING THE SPECIFICATIONS PHASE
 *Structural walkthroughs

TESTING DURING THE DESIGN PHASE
 *Design inspections

TESTING DURING THE IMPLEMENTATION PHASE
 *Equivalence testing & boundary analysis
 *Functional testing
 *Statement, branch, and path coverage
 *Complexity
 *Correctness

TESTING DURING THE INTEGRATION PHASE
 *Module testing
 *Acceptance testing
 *Hardware testing

TESTING DURING THE OPERATIONS PHASE
 *Corrective testing
 *Regression testing

FIG. 6.5. Testing and the life-cycle model.

a very critical phase of the software development life cycle. Techniques and tools that effectively represent requirements in a format that results in a fault-free product are essential to a good program design. A description of these tools and their evaluation form the subject of this section.

A design methodology is an artificial language that enables the programer to describe a particular type of problem at a conceptual, rather than implementation, level. The tools (i.e., pseudo code, Warnier Orr diagrams, Hypo-Charts) that are derived from the design methodology allow the programer to be precise about which parts of the program are program specific and which parts address a more general design plan. Programers who use design methodologies produce programs that contain fewer errors and run faster because they are forced to divide the problem into smaller modules that are easier to design, code, and debug. Such a methodology permits the designer to cooperatively develop systems using a shared language of architecture constructs, rather than a set of problem-specific primitives.

Different design methodologies tend to emphasize different aspects of the programing process. For example, some design methodologies are very effective for showing data and the relationship among data items, whereas other methodologies stress data flow or program functionality (Burch, 1992; Shach, 1990). Two design techniques that highlight different aspects of the design process are Petri nets and entity-relationship (ER) diagrams. Petri nets have proven extremely useful for describing real-time systems because they describe the flow of data throughout the system. In contrast, ER diagrams are effective for representing the object-oriented programing paradigm because they show data and the relationships among data. The connection between the specification language and the problem description can be very critical, as shown later.

Petri nets are abstract, formal models of information flow that look very similar to directed graphs (Schach, 1990). As illustrated in Fig. 6.6, nodes are used to represent completion of events, and arcs represent transitions from one event to another (Peterson, 1980). Petri nets have been successfully used for problems relating to parallel computation (Ghezzi & Mandrioli, 1987; Miller, 1979), multiprocessing (Agerwala, 1979; Guha, Lang, & Bassiouni, 1987), knowledge representation (Jantzen, 1980), and human information processing (Schumacker & Geiser, 1978). Although Petri nets can be used to represent descriptive data, they are more suited to describing information flow. For this reason, Petri nets are useful for specifying real-time systems, with timing issues being critical.

As illustrated in Fig. 6.7, an entity-relationship (ER) diagram consists of nodes that represent entities and arcs that represent the relationship between two entities. ER diagrams were first introduced by Chen (1976), who used them to describe the entity-relationship data base model. As such, ER diagrams are more appropriate for describing data and the relationships among data. More recently, ER diagrams have been proposed as a design language for knowledge-base

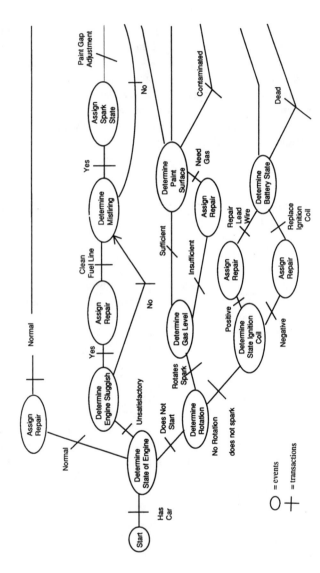

FIG. 6.6. Example of Petri net representation of an expert system for troubleshooting and repairing an automobile.

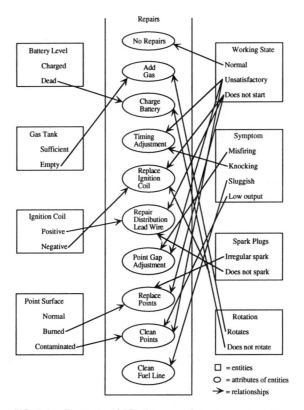

FIG. 6.7. Example of ER diagram of an expert system.

development (Addis, 1985; Swigger & Brazile, 1988) and have proven effective for describing object-oriented systems (Boehm-Davis, Holt, Schultz, & Stanley, 1986).

Design Comparisons.   Many other formal techniques have been proposed. For example, Anna (Luckham & von Henke, 1985) is a formal specification language for Ada, whereas Refine (Smith, Kotick, & Westfold, 1985) and Gist (Balzer, 1985) are used to describe knowledge-based systems. The structural analysis and design technique (SADT) has had great success in specifying a wide variety of products, especially complex, large-scale military projects (Ross, 1985). Research has recently discovered that the decision to use a particular design technique for a specific project depends on the problem that needs to be solved (Pennington & Nicolich, 1991). Boehm-Davis and Ross (1984), Boehm-Davis et al. (1986), Boehm-Davis and Fregly (1983) performed a number of studies that were aimed at determining the effect of using different design/documentation formats in a variety of comprehension, coding, verification, and modification tasks. Boehm-Davis and Ross (1984) first determined that performance on a limited set of software tasks was linked to documentation type. Boehm-Davis and Fregly (1983) expanded the earlier study by comparing different documentation formats such as PDL, abbreviated English, and Petri nets and found that performance scores varied as a function of documentation type. Finally, Boehm-Davis et al. (1986) asked experienced programers to modify several different types of programs in combination with several different types of design tools. The authors concluded that different design/documentation formats did indeed affect both design time and problem solution time and that the differences could be attributed to the different types of formats.

Similar experiments have examined different design tools used to program expert systems. Swigger and Brazile (1989, 1990) found that programers who used a design tool performed significantly better (i.e., modified the program in less time) on modification tasks than programers who did not use a design tool. Results also suggested that there were differences between different types of design tools, and that different design tools affected different types of programing tasks.

What seems to be important for the development of both conventional software as well as expert systems is that the design technique provide a uniform representation and organization of the more general problem description. This should also be true of design tools that are used to develop educational software and other types of applications. Thus, it appears to be necessary to identify a software product as belonging to a specific class or type of problem and then use the design techniques that best represent the problem type. This type of classification relates to both the problem's domain knowledge and the programing techniques used to solve the problem.

AUTOMATED PROGRAMING TOOLS

The idea that domain knowledge and programing knowledge are equally important for the construction of good software has had a major impact on the development of automated programing tools. Although the concept of a total automatic programing environment remains a fantasy, there are several recent developments that bring the fantasy closer to reality. These types of tools are known as computer-aided software engineering (CASE) products.

In the past, the terms *automatic programming* and *CASE* were applied to any type of programing tool that automated any part of the program life cycle (Balzer, 1985). It is only recently that companies have developed products that, through a successive series of steps, are able to transform specifications into executable source code. The transformation process is by no means fully automated; human intervention is required in deciding which transformations to

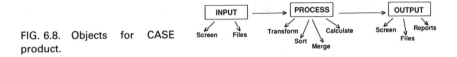

FIG. 6.8. Objects for CASE product.

apply, and precisely where to apply the transformations. Underlying these successful products is a model of programing as well as a model of the domain.

One approach to a CASE tool is to conceptualize the programing model as consisting of inputs, processes, and outputs (Fig. 6.8). Input objects include both screen and file objects. Output objects include screen and file objects as well as report objects. Then, depending on the application, process objects (e.g., sort, sequence, retrieve, etc.) are used to transform the input–output objects. The CASE tool creates the different objects (i.e., screen, report, sort, etc., objects) by asking the programer to supply both domain-specific and programing knowledge. For example, the CASE tool creates a specific report object by asking the programer to provide the format of the report, the heading, the names of the specific data items to be processed, the primary control break, and so on. Thus far, CASE tools have been developed only for restrictive domains such as business applications (Frenkel, 1985), database problems (Kaiser, Feiler, & Popovich, 1988), and data analysis problems (Balzer, 1985).

Another approach to automated programing is to build specialized tools for a specific programing language. Such tools can handle much of the drudge work of programing, leaving the creative work to the human programer. Powerful debuggers, intelligent editors, and elaborate programing environments are included under this approach. It has been argued that the use of such tools is, in itself, a sufficient condition for increased programer productivity and effectiveness (Barstow, Shrobe, & Sandewall, 1984). For example, there are powerful debuggers for the C programing language that optimize code, provide powerful debug messages, and suggest effective programing styles (e.g., Borland's C and C++).

A third type of computer tool focuses on the problem of supporting team programing and large product development projects. Computer-supported cooperative work (CSCW) is a research area that includes the development of computer systems that support group design activities. For example, the software technology program at Microelectronics and Computer Technology Consortium is working on the problem of issue-based information systems (IBIS), which will help software designers by supporting structured collective conversations through planning (Conklin, 1986). In a similar manner, this author has recently built and evaluated an intelligent interface to support computer-supported cooperative problem solving. The system serves as a testbed for investigating tools that people use while engaged in technical, cooperative tasks such as working on large programing projects (Swigger & Thomas, 1992). Each of the on-line tools

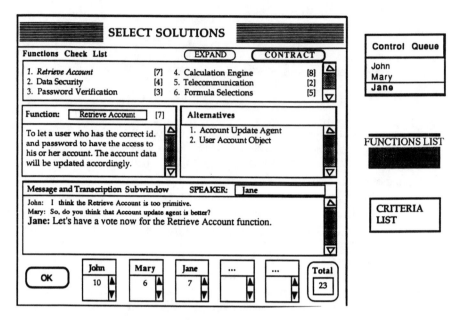

FIG. 6.9.   Example screen for computer-supported cooperative environment.

represents and is linked to a requirement for successful communication (Fig. 6.9). For example, following an examination of how to build common vocabularies, an on-line tool was created that enhances this requirement. Thus, the system is designed to test a "theory of communication" stating that effective cooperative problem solving is dependent on effective communication, which, in turn, consists of common vocabularies, syntax, objectives, and so on. If the communication model is correct, then the on-line tools should enhance communication performance.

Whether in the form of a CASE tool, a powerful debugger, or a CSCW tool, recent advances in programing incorporate models of programing as well as models of the domain. As research indicates (Pennington, 1987), computer programing is a highly complex task with many components. It is often compared to other types of tasks in an attempt to understand its underlying processes. For example, programs have been compared to texts (Atwood & Ramsey, 1978; Thomas & Zweben, 1986). As such, they are described as having organization and structure, and programers are said to have general schemata that guide encoding, representation, and retrieval of the program-as-text (e.g., Rumelhart, 1977). Programing has also been compared to expert skills such as playing chess or Go, and diagnosing faulty electric circuits. This particular analogy focuses attention on the potentially large stores of specific programing patterns that have been learned through extended practice (Barstow et al., 1984; Chase & Simon, 1973; Chi, Feltovich, & Glaser, 1981). Finally, programing has been analyzed as a planning and problem-solving task that utilizes some general strategies such as problem decomposition and reformulation (Miller & Goldstein, 1977; Newell & Simon, 1972; Soloway, Pinto, Letovsky, Littman, & Lampert, 1988). The suggestions for program planning have been much closer to the computer scientist's view of orderly, top-down structured programing than the psychological literature would lead us to expect.

It should be noted that the previous analogies are not necessarily contradictory. Programing-as-text focuses on comprehension and memory and, as such, implies that the programer works backward from program to interpretation. Programing-as-planning focuses on program construction and implies that the programer works forward from problem analysis to the program. Programing-as-expert skill focuses on the organization of knowledge specific to the programing domain that is clearly implicated in both comprehension and construction of programs. The challenge is to develop programing tools and software evaluation techniques that reflect the human aspects of programing as well as the specific problem domain. As such, the software tools and the software evaluation techniques need to incorporate models of programing and models of the domain.

SUMMARY AND IMPLICATIONS
FOR FURTHER RESEARCH

One goal in extensively reviewing the existing studies on software evaluation and evaluation of software engineering has been to identify important themes that pertain to programing as well as technology assessment. Across evaluation tasks, a recurring question concerned the existence and nature of meaningful measures with which to evaluate software and software engineering practices. Several measures were examined, and a wide variety of evaluation techniques and formulations were found that address different aspects of the software development cycle.

The chapter first distinguished between software and software engineering and stated that the difference related to their use and function: specific code that solves a domain-specific problem versus a model or methodology for general program development. As a result of this distinction, it was possible to discuss code evaluation as opposed to the evaluation of a programing methodology. A second distinction concerned the model of programing that each of these two areas measures; error-free code versus general problem solving. Although such distinctions exist, it was noted that most programs are now written as part of larger systems. As such, software and software engineering involve the use of general problem-solving strategies as well as specific domain knowledge. Thus, a second recurring set of questions concerned the definition of software (or software engineering) and, indirectly, the definition of programing: (a) Is programing a series of successive transformations of the external problem domain into a representation in the programing language? (b) Can a program be evaluated separate from either its domain or programing techniques? (c) Are there fundamental structural components of programing that exist?

This review also documented that conflicting evidence exists on software and software engineering evaluation. Although software evaluation studies have not demonstrated the existence of a single set of principles for software evaluation, there are several software measures that can be applied to other areas of technology assessment. Similarly, software engineering studies addressing larger issues of software design fail to report a single design methodology that results in the production of quality software over time and for every type of application. Yet, the idea of using different tests for different phases of the software development life cycle is a major lesson to be learned from these studies. A second lesson seems to be that every development group must decide what type of evaluation procedure is appropriate for a particular problem type.

In considering the various topics in this chapter, it has been noted that it is difficult to draw definitive lines between evaluation of software and evaluation of software engineering. It has been argued that an understanding of a model of programing is closely related to both the construction of effective software and a programer's systematic approach to a problem. A model of programing involves the cognitive representation of particular programs at the surface level of the code, at a deeper structural level, and at an interpretive level. It also involves a similar representation of the knowledge of the application and a deep understanding of how it will be used by the client.

One reason for focusing on developing models of programing and then using these models to derive performance measures for software is that this method has implications for the development of programing aids. One can imagine, for example, a programing tool that is capable of transforming different representations at different levels of abstraction. The tool would be able to analyze domain-specific information and use this knowledge to suggest possible programing strategies. A different scenario would entail using a programing tool to design screens, write procedures, and develop data structures. The tool would interrupt the programer only when it identified an error. A third type of tool would enable programers located in different cities and countries to work together on a single programing project. Such a tool would allow team members to exchange code, documents, and design information in an effort to create a useful, robust, reliable, and bug-free program. Regardless of which version of the future one chooses, the development of an effective programing tool depends on a clear understanding of the programing process as well as the application domain.

Other reasons for studying models is that they have implications for the development of measures of program assessment. Current programing measures are inadequate to evaluate large programing projects. Testing all paths or constructing sufficient test cases are impractical for current software development projects. Although every piece of software must produce correct results; it must also be acceptable to the user. Therefore, it is important to consider the human aspects of software development to determine issues of complexity, maintainability, and usability. It has been documented that programer productivity increases, complexity decreases, and program performance increases, when software engineers use "good" programing practices (Boehm & Papaccio, 1988). Being able to define "good" programing practices in terms of a model of human performance should also improve productivity and performance.

There is a final lesson in this analysis that has relevance to the general issue of technology assessment. Once software is transferred from the programer to the user, the question of software evaluation or even evaluation of software engineering is no longer relevant. At this point, ask the following question: Is the underlying "model" of pedagogy, communication, explanation, learning, and so on, as represented in the computer program, correct and effective? As a program advances to its final stages of development, it ceases to be a program and becomes a model of human performance. Therefore, product testing and evaluation should concentrate on the model and not the program.

The intention of this chapter has been to review the existing practices in software engineering for program evaluation, to identify some recurring questions, and to suggest some implications of human behavior for software evaluation and for technology assessment in general. There are clearly other avenues of research that might be pursued productively in the study of assessment of software practices. However, these avenues will necessitate an understanding of the human aspects of problem solving and the way that these aspects interact with specific domains.

ACKNOWLEDGMENTS

This research was supported, in part, with funds from the National Science Foundation under contract IRI-9109547 and from the Ernst and Young Foundation.

The research reported in this chapter was supported in part by a contract from the Devense Advanced Research Projects Agency (DARPA), administered by the Office of Naval Research (ONR), to the UCLA Center for the Study of Evaluation/Center for Technology Assessment. However, the opinions expressed do not necessarily reflect the positions of DARPA or ONR, and no official endorsement by either organization should be inferred.

REFERENCES

Addis, T. R. (1985). *Designing knowledge-based systems.* Englewood Cliffs, NJ: Prentice-Hall.
Agerwala, T. (1979). Putting petri nets to work. *Computer,* December, 85–94.
Atwood, M. E., & Ramsey, H. R. (1978). Cognitive structure in the comprehension and memory of

computer programs: An investigation of computer program debugging (ARI Tech. Rep. No. TR-78-A210). Englewood, CO: Science Applications.

Balzer, R. (1985). A 15 year perspective on automatic programming. *IEEE Transactions on Software Engineering, SE-11*, 1257–1268.

Barstow, D. R., Shrobe, H. E., & Sandewall, E. (Eds.). (1984). *Interactive programming environments.* New York: McGraw-Hill.

Basili, B. R., & Hutchens, D. H. (1983). An empirical study of a syntactic complexity family. *IEEE Transactions on Software Engineering, SE-9*, 664–672.

Basili, B. R., & Selby, R. W. (1987). Comparing the effectiveness of software testing strategies. *IEEE Transactions of Software Engineering, SE-13*, 1278–1296.

Boehm, B., & Papaccio, P. (1988). Understanding and controlling software costs. *IEEE Transactions on Software Engineering, 14*, 1462–1477.

Boehm-Davis, D. A., & Fregly, A. (1983). Documentation of concurrent programs. In A. Janda (Ed.), *CHI '83 Conference Proceedings* (pp. 256–261). Boston:

Boehm-Davis, D. A., Holt, R. W., Schultz, A. C., & Stanley, P. (1986). The role of program structure in software maintenance (Tech. Rep. No. Tr-86-gmu-po1). Fairfax, VA: George Mason University.

Boehm-Davis, D. A., & Ross, L. (1984). Approaches to structuring the software development process (Tech. Rep. No. Tr-84-B1v-1). Arlington, VA: General Electric Company.

Brazile, R., & Swigger, K. M. (1990). Criteria for selecting different modeling techniques. *Proceedings of the International Conference on Modelling and Simulation*, pp. 24–28. International Association of Science and Technology for Development (IASTED), Anaheim, CA.

Burch, J. (1992). *Systems analysis, design, and implementation.* Boston: Boyd & Fraser.

Chase, W. G., & Simon, H. A. (1973). Perception in chess. *Cognitive Psychology, 4*, 55–81.

Chen, P. (1976). The entity-relationship model—toward a unified view of data. *ACM TODS, 1*, 9–36.

Chi, M.T.H., Feltovich, P. J., & Glaser, R. (1981). Categorization and representation of physics problems by experts and novices. *Cognitive Science, 5*, 121–152.

Conklin, J. (1986). *A theory and tool for coordination of design conversations* (Tech. Rep. No. STP-236-86). Austin, TX: Microelectronics and Computer Technology Consortium.

Dijkstra, E. W. (1976). *A discipline of programming.* Englewood Cliffs, NJ: Prentice-Hall.

Dijkstra, E. W. (1990). *Predicate calculus and program semantics.* New York: Springer-Verlag.

Fagan, M. E. (1976). Design and code inspections to reduce errors in program development. *IBM Systems Journal, 15*, 182–211.

Frenkel, K. A. (1985). Toward automating the software-development Cycle. *Communications of the ACM, 28*, 578–589.

Gelperin, D., & Hetzel B. (1988). The growth of software testing. *Communications of the ACM, 32*, 687–695.

Ghezzi, C., & Mandrioli, D. (1987). On eclecticism in specifications: A case study centered around Petri nets. *Proceedings of the fourth international workshop on software specification and design* (pp. 216–224). Washington, DC: IEEE Press.

Goodenough, J. B. (1979). A survey of program testing issues. In P. Wenger (Ed.), *Directions in software technology* (pp. 319–340). Cambridge, MA: MIT Press.

Goodenough, J. B., & Gerhart, S. L. (1975). Toward a theory of test data selection. *IEEE Transactions on Software Engineering, SE-1*, 156–173.

Guha, R., Lang, S., & Bassiouni, M. (1987). Software specification and design using Petri nets. *Proceedings of the fourth international workshop on software specification* (pp. 225–230). Washington, DC: IEEE Press.

Halstead, M. H. (1977). *Elements of software science.* New York: Elsevier North-Holland.

Howden, W. E. (1987). *Functional program testing and analysis.* New York: McGraw-Hill.

Jantzen, M. (Ed.). (1980). Structured representation of knowledge by Petri nets as an aid for teaching and research. *Lecture notes in computer science.* Berlin: Springer-Verlag.

Kaiser, G. E., Feiler, P. H., & Popovich, S. S. (1988). Intelligent assistance for software development and maintenance. *IEEE Software, 5,* 40–49.

Leavenworth, B. (1970). Review #19420. *Computing Reviews, 11,* 396–397.

Luckham, D. C., & Von Henke, F. W. (1985). An overview of Anna, a specification language for Ada. *IEEE Software, 2,* 9–22.

Manna, A. (1974). *Mathematical theory of computation.* New York: McGraw-Hill.

McCabe, T. J. (1976). A complexity measure. *IEEE Transactions on Software Engineering, SE-2,* 308–320.

Miller, G. A. (1956). The magical number seven, plus or minus two: Some limits on our capacity for processing information. *Psychological Review, 63,* 81–97.

Miller, M., & Goldstein, I. (1977). Structured planning and debugging. *Proceedings of the Fifth International Joint Conference on Artificial Intelligence.* Los Altos, CA: Morgan Kaufman.

Miller, P. (1979). A methodology for the design and implementation of communication protocols. *IEEE Transactions on Communication, 24,* 614–621.

Myers, G. J. (1978). A controlled experiment in program testing and code walkthroughs/inspections. *Communications of the ACM, 21,* 760–768.

Newell, A., & Simon, H. (1972). *Human problem solving.* New York: Prentice-Hall.

Pennington, N. (1987). Comprehension strategies in programming. In G. Olsen & E. Soloway (Eds.), *Empirical studies of programming: Second workshop* (pp. 100–114). Norwood, NJ: Ablex.

Pennington, N., & Nicolich, R. (1991). Transfer of training between programming subtasks: Is knowledge really use specific? In J. Koenemann-Belliveau, T. Moher, & S. Robertson (Eds.), *Empirical studies of programmers: Fourth workshop* (pp. 156–176). Norwood, NJ: Ablex.

Perry, W. W. (1983). *A structured approach to systems testing.* Englewood Cliffs, NJ: Prentice-Hall.

Peterson, J. L. (1980). *Petri net theory and the modeling of systems.* Englewood Cliffs, NJ: Prentice-Hall.

Rapps, S., & Weyuker, E. J. (1985). Selecting software test data using data flow information. *IEEE Transactions of Software Engineering, SE-11,* 367–375.

Ross, D. (1985). Applications and extensions of SADT. *IEEE Computer, 18,* 25–34.

Rothenberg, J. (1990). *Prototyping as modeling: What is being modeled?* (Rep. No. N-3191-DARPA) Washington, DC: The RAND Corporation.

Royce, W. W. (1970). Managing the development of large software systems: Concepts and techniques. *Proceedings of WestCon,* 50–55.

Rumelhart, D. E. (1977). Understanding and summarizing brief stories. In D. LaBerge & S. J. Samuels (Eds.), *Basic processing in reading: Perception and comprehension* (pp. 265–303). Hillsdale, NJ: Lawrence Erlbaum Associates.

Schach, S. R. (1990). *Software engineering.* Homewood, IL: Aksen Associates.

Schumacker, W., & Geiser, G. (1978). Petri nets as a modeling tool for discrete concurrent tasks of the human operator. In *14th Annual Conference on Manual Control, NASA Conference Publication, 2060,* 161–197.

Shneiderman, B. (1980). *Software psychology: Human factors in computer and information systems.* Cambridge, MA: Winthrop Publishers.

Smith, D. R., Kotik, G. B., & Wesfold, S. J. (1985). Research on knowledge-based software environments at the Kestrel Institute. *IEEE Transactions of Software Engineering, 11,* 1278–1295.

Soloway, E., Pinto, J., Letovsky, S., Littman, D., & Lampert, R. (1988). Designing documentation to compensate for delocalized plans. *Communications of the ACM, 31,* 1259–1267.

Swigger, K. M., & Brazile, R. (1988). An architecture for an intelligent tutoring system based on entity-relationship data model. In C. Frasson (Ed.), *Proceedings of the International Conferences on Intelligent Tutoring Systems* (pp. 258–263). Montreal, Canada: University of Montreal.

Swigger, K. M., & Brazile, R. (1989). Experimental comparison of design/documentation formats for expert systems. *International Journal of Man-Machine Studies, 31,* 47–60.

Swigger, K. M., & Brazile, R. (1990). An empirical study of the effects of design/documentation formats on expert system modifiability. In J. Koenemann-Belliveau, T. Moher, & S. Robertson (Eds.), *Empirical studies of programmers: Fourth workshop* (pp. 210–216). Norwood, NJ: Ablex.

Swigger, K. M., & Thomas, T. (1992). A computer-supported cooperative problem solving environment for examining communication effectiveness. *Proceedings of International Conference on Intelligent and Cooperative Information Systems* (pp. 98–103). M. Huhns, M. P. Papazoglou, & G. Schlageter (Eds.). Washington, DC: IEEE Press.

Takahashi, M., & Kamayachi, Y. (1985). An empirical study of a model for program error prediction. *Proceedings of the Eighth International Conference on Software Engineering,* 330–336. Washington, DC: IEEE Press.

Thomas, M., & Zweben, S. (1986). The effects of program-dependent and program-independent deletions on software cloze tests. In E. Soloway & S. Iyenger (Eds.), *Empirical studies of programmers* (pp. 138–152). Norwood, NJ: Ablex.

Woodward, M. R., Hedley, C., & Hennell, M. A. (1980). Experience with path analysis and testing of programs. *IEEE Transactions on Software Engineering, SE-6,* 278–286.

7

An Evaluation Model for Investigating the Impact of Innovative Educational Technology

Robert J. Seidel
Ray S. Perez
U.S. Army Research Institute

We have worked with and around the application and evaluation of computers in education and training for over 30 years and have witnessed few eternal truths.
Consider the following:

1. Repeatedly, we have observed that a wave of blind enthusiasm greeted each new technology. Be it hardware or software, mainframe or personal computers, it was advertised with such overstatement and promise that the naive training or educational users saw the innovation as a technological savior for all (or a good portion) of the training and/or educational system deficiencies.

2. The person or groups who pushed for obtaining the technology were surely the same people who paid for it (Seidel et al., 1978).

3. The purpose for the application of the technology was frequently stated in vague generalities—to improve instruction, to aid the teacher, or to train developers or administrators (Seidel, 1974).

4. The evaluation of the technology as perceived by the persons or groups who are pushing it rarely see the value of the evaluative process. Thus, the participants of the "technology push" believe that the technology a priori is of value and it is better than what proceeded it (O'Neil & Baker, 1991).

5. As a corollary to the previous points, the technology rarely was evaluated in terms of reasons for which it was obtained, and never was consideration given to exploiting the technology for added value not originally perceived and yet for which it might be aptly suited (Wagner & Seidel, 1978).

6. Perhaps it goes without saying, but the reasons for not institutionalizing computer technology as a natural, integrated component of the education (train-

ing) system has never been the fault of the technology. It has been a result of failure to evaluate the technology in the proper organizational context (school environments, surrounding community, administrator, bill payers, instructors, required facilities, or, of course, students).

This chapter attempts to take these verities into account as we present a model for evaluating technology in training and education. This chapter is divided into two parts. The first part is a selective review of studies on the use and evaluation in schools, military and job settings covering issues and methods that highlight the previous points. The second part presents a discussion of current methods and issues in evaluating the impact of computer technology and a description of a multilevel and multidimensional model for assessing the value of using technology for education and training. We hope the latter model can provide some useful suggestions to the next generation of potential innovators; guidance perhaps, that can facilitate, not just the introduction of computers and other advanced technologies into training and education, but the integration of them as natural elements of any instructional system.

PART I: USE AND EVALUATION RESULTS OF COMPUTER TECHNOLOGY

To determine the impact of technology on the way we learn, work, and play has been an objective of research for some time. The most obvious question to be asked of these technologies is do they help? Perhaps the more difficult but important research question is how do they help? The research on the impact of computer technology in educational and training environments has historically, in part, been concerned with assessing its impact on the environment in terms of student achievement and cost factors. More recently, the emphasis has changed to discovering how technology is actually used, how it works, and more importantly, how to describe the fundamental changes in the nature of the environment and the learning process of the users/students as a result of the introduction of the technology.

Although the introduction of the first microcomputer kit to the U.S. consumer occurred in 1975 (Tobias, 1984), in that short span of time the computer has had a significant impact on the way we learn and think (Fletcher, Hawley, & Piele, 1990; Liao, 1992). The power of the computer as a tool for number crunching, word processing, and teaching has become a central topic for discussion among educators and psychologists (Pea, 1985). This, along with the increasing availability of powerful and relatively inexpensive personal computers and software has brought the cost of computers within the budget constraints of most, if not all, school districts. The promise of computer technology to education has been to provide the technological seed for revolutionizing training/education. The form of use could be anything from simple drill and practice computer-based instruction (CBI) to active problem solving in complex microworlds or the creation of virtual reality environments.

Use of Computer Technology in Schools

In 1981 there were between 1.2 and 1.7 million computers in U.S. public schools. Over 95% of all elementary and secondary schools now have at least one computer intended for instructional use, compared to 18% in 1981. This provides one computer for every 30 children enrolled in U.S. public schools (Office of Technology Assessment, 1988, chap. 1). Becker (1991) estimated that there are roughly 2.4 million microcomputers and computer terminals in U.S. schools. Video technology has also experienced a phenomenal growth in public schools similar to that of computers, with 91% of all public schools having some form of video technology (videocassette records, VCRs) for instructional purposes (Office of Technology Assessment, 1988). These figures illustrate the increasing rate and magnitude of the introduction of computers in educational settings. The support from the two principal organizations responsible for research and development (R&D) of educational technology has come from the National Science Foundation (NSF) and Department of Defense (DOD). NSF's R&D efforts have primarily been directed at influencing innovative technology development and use in science and mathematics education and more recently the use of computer to enhance the development of these basic skills. The Department of Education R&D efforts, on the other hand, have been spent in developing and supporting educational television and public broadcasting and in increasing access to technology by enabling schools to purchase hardware and software. The Department of Education (DOE) also has research efforts aimed at the use of computer technology for the handicapped. To a lesser extent DOE has been concerned with the expansion of the base of knowledge and innovations. DOE efforts have been accomplished by a bottom-up approach where schools were provided funds to purchase hardware and software (televisions, videocassette, computers, and software); thus, as the number of computers increased in schools, their use would expand, driving the development of new products (courseware) to supply the school market (Office of Technology Assessment, 1988, chap. 1).

How has the usage of computers in U.S. schools changed in the last 10 years? In 1985 the "Second National Survey of Instructional Uses of School Computers" found that teachers in elementary schools rarely used computers as a regular means of providing students with instruction or practice in traditional school subjects. Rather, computers were used primarily for enrichment by secondary level school teachers and to teach computer literacy and computer programming classes. The focus of these computer programming classes were on the teaching of the syntax of BASIC, rather than on the more general principles of programing and problem solving. However, since 1985, the usage of computers in U.S. schools has slightly changed to the systematic and regular use of computers to provide practice of basic skills in elementary school laboratories. In middle and high schools there has been an increased emphasis on the use of computers as productivity tools for the expression of ideas (e.g., wordprocessing programs), the recording (e.g., data base programs), and analyzing of information (e.g., spreadsheets).

Although we have witnessed a dramatic increase in the number of computers in U.S. schools we still have not achieved the goal of providing access to computers for every student in our schools. In fact, quite the contrary, Becker (1991) found that "the number of students in one school who can simultaneously use computers remains small compared to the number of students who can be simultaneously served by teachers, blackboards, books, worksheets, manipulative materials, or even film and video projectors " (p.330).

Use and Evaluation of Computer Technology in the Military.   The use of computer technology for training and education in the military has had a history of growth similar to that of as NSF and DOE. The DOD has played a major role in the development of computer technology and its applications to education and training. DOD in fiscal year 1990 spent approximately $289 million on R&D technologies in training/education, and a major portion ($129 million) of this research was spent on the development of simulators, whereas civilian spending during the same year on R&D technology for K–12 education was estimated at $30 million. DOD uses R&D strategies similar to those found in civilian training and education. DOD R&D investment in technology has ranged from basic cognitive science investigators to applied development of course materials and electronic teaching machines (e.g., CAI). The result of this investment has been the development of training applications (courseware) of direct utility to the military (Fletcher & Rockway, 1986). Military training differs from most civilian activities in at least one respect-the number of those trained. The military trains 200,000 students on any given day. In 1987 DOD spent about $18 million on formal residential training *Military Manpower Training Report for FY 1987:* Vol. 4 *Force Readiness Report.* This does not take into account the costs for training that takes place in the field or on the job. So for example, under the Navy's old training system, saving an hour of instruction using computer-assisted instruction and computer-managed instruction in the Navy's 14-week basic electricity course would save 20,000 person hours per year in student time alone, not counting the savings in instructor time and upkeep of facilities which is a significant impact (Halff, Holland, & Hutchins, 1986).

Use and Evaluation of Computer Training in the Workplace.   Unlike the military and education, the primary driving force for use of computers in industry has been for increasing efficiency and productivity through office automation.

The introduction of personal computers (PCs) in the early 1980s offered the promise of increased efficiency and productivity to decision makers and managers in industry. However, it has been difficult to assess the impact of computer technology on the way we work. The one issue probably most frequently misunderstood is the cost-benefit or cost-effective attempt at evaluation of computers in the workplace.

This cost-benefit analysis approach attempts to quantify the impact of the introduction of the innovation (PC) on the workplace. The cost-benefit ratio is generally how much benefit is derived from the amount of investment one makes. One such measure is worker productivity. The algorithm generally associated with this metric is cost over increased unit of output production (Orlansky & String, 1979; Seidel & Wagner, 1979).

During the 1980s, U.S. companies in the services sector industry doubled the amount of money spent on high-tech "productivity tools." They spent, on the average, $9,000 per employee. In spite of this investment in computer hardware and software, they noted that the output of these workers only increased by .02% per year (Boroughs, Black, Ferguson, & Pasternak, 1990). Although, the output measures of productivity in this study were subjective and difficult to quantify because of the nature of the products of the service industry; that is, it's not a tangible thing or widget, the results of this survey still were unexpected. The necessity of finding and using objective and quantifiable measures become more important, because the service sector of U.S. industry now accounts for more than 70% of U.S. companies. This "productivity paradox" was observed during the 1980s. And despite increased investment in high-tech office equipment, white-collar productivity has remained flat (*U.S. News & World Report,* 1991). The reason for this observed paradox is twofold.

First, it is the way people use the technology rather than the technology itself that increases production. Most people are not trained to use the efficiencies of the computer. An example of this is an anecdotal account given by Boroughs et al., 1990). They described an employee of an international sporting goods manufacturer who would laboriously enter a department Roladex of phone numbers into her computer. Her goal was not to save labor with fast computer searches for names and number. Instead, she printed out the computerized results onto new Roladex cards that she then distributed to her sales staff, who then would flip through the cards for phone numbers of clients. Boroughs et al. (1990) suggested that one remedy is to train people to rethink their work in order to take advantage of the computers efficiencies. Computers will not improve productivity if people are only automating manual processes.

Second, PCs were originally not purchased to be integrated into the workplace. Because most firms did not purchase computers in a systematic way, they did not plan for integration. Therefore, most PCs in the workplace cannot talk to each other or existing mainframes. They lack the ability to access much needed information (data bases) stored on mainframe computers. Strassman (1990) argued against the notion of individual productivity in the workplace; rather, he suggested that it should be replaced with *organizational* productivity. People work collaboratively in industry. The services and goods they produce are not the result of one individual but the combined efforts of a number of individuals working together. More often than not, PCs were purchased as isolated machines for use by one person. Such a computer is not going to improve the flow of information around the office and is not going to contribute organizational productivity. Strassman argued that the solution to this problem is to apply the technology of computer networking that will improve the flow of information around the office and thereby increase productivity. The impact of computers on organizational productivity has been studied by several researchers (De Sanctis & Gallupe, 1987, Hiltz, Johnson, & Turnoff, 1986, Keisler, Siegel, & McGuire, 1984). They studied the impact of computers on the productivity of work groups, using a number of different computer systems: (a) simple wordprocessing programs or accounting programs; (b) complex collaborative writing/editing programs; (c) electronic mail, bulletin boards, or meeting rooms; and (d) group decision support systems. The clearest evidence that they found involved the impact of electronic communication; it appears to affect work groups by reducing overall miscommunication, equalizing participation levels, weakening status systems, emphasizing informational rather than normative influences, and encourages certain forms of deviance. They found little evidence of increased productivity for work groups as a result of the introduction of computers.

The same issues that confront use of computer technology confront those who wish to evaluate its impact: The computer has the flexibility to be employed by the users in multiple ways and at multiple levels. Here again, how you evaluate the impact of the computer on the workplace will be based on the users' purposes and the organizational structure to help or hinder these purposes.

The Computer as a Cognitive Tool.   In recent years the application and the use of computer technology has evolved to a focus on the development of software tools to be used in industry and in classroom and for more specific applications such as the development of training and educational materials (see Perez & Seidel, 1990, for a discussion of tools for training development). Within this context we speak of two kinds of tools. The first type is software designed to help people perform specific tasks better and faster (Hawkins & Kurland, 1987). Examples of these tools are word processors (e.g., word perfect), data base management systems, spreadsheet programs (e.g., Lotus 1,2,3), and graphic and text editors. The assumption made about the intended users of these tools is that they already know how to perform the specific task for which the tool is designed. The second type of tools are "intelligent tools," including decision aids (e.g., expert systems) and intelligent job aids or advisors. These tools are designed not only to help people perform tasks more efficiently and effectively, but also to supplement the user's decision-making capabilities by providing them with a knowledge base of expertise (Barr & Feigenbaum, 1982; Perez & Seidel, 1990). In the case of this latter type of tool, the designers have anticipated that the intended users may or may not be proficient at the task that the tools were designed for and have included tutorials or on-line help systems.

To summarize our discussion of computer tools, consider Hunter's (1988) list of these tools in education or training environments, as opposed to the use of the computer as a teaching machine:

- They are under the control of the person performing the task,
- They can be used to accomplish a variety of goals and tasks for a variety of purposes,
- The users determine the goals and purposes of the task at hand,
- The skills and intentions of the users determine the quality and nature of the process and product.

Whether these tools are implemented and used in educational or training settings, the problems of how to measure and assess their impact are essentially similar to that of evaluating the impact of computers.

Computer Tools for the Automation of a Job: Use and Evaluation.   In this section we describe another type of cognitive tool, automated job aids, and discuss one prototype automated job aid, the U.S. Army's Automated Systems Approach to Training (ASAT).

An automated job aid can be defined as a specific software program designed such that it embodies the procedures and processes that are part of, and necessary for, the performance of a specific job/task. This type of software tool is designed to help people perform jobs/tasks better. Automated job aids differ along many important dimensions from information-managing software tools, such as word-processors, data base management systems, spreadsheet programs, and graphic editors. One essential difference is the user's job functions. Generally, they are customized software programs for use on a specific set of tasks, within a specific work setting, and with a specific user in mind. These programs are *not* flexible generic tools designed for a variety of goals, tasks, and purposes. Because these software tools embody the procedures and processes that are part of and necessary to the performance of a particular application, their use is limited to a specific context. They can only be used to accomplish a limited set of goals and tasks and the purposes are limited to the job, that is, the goals and purposes of the tool are determined by the job/task.

The use of these tools are highly dependent on the users' skills, experience, knowledge, and intentions. The quality and nature of the process and product of those tools to a great extent is determined by the users' characteristics more than the more generic information managing type of tools. An example of such a tool is the U.S. Army's Automated Systems Approach to Training (ASAT), (Perez, Gregory, & Minionis, 1992). ASAT was designed to automate a portion of the procedures and processes of the curriculum development system used by the military to produce training. These procedures and processes are known as the Systems Approach to Training (SAT). The automation of the SAT procedures and processes resulted in the automation of a significant portion of the military training developer's job. More specifically, the initial version of ASAT auto-mates the analysis and design phases of the SAT process. ASAT aids the training developers in conducting a front-end analysis (job/task analysis) for both groups and individuals; helps them in identifying collective and individual critical tasks; and provides tools for the design and development of a mission training plan, drills, and lessons. It accomplishes this by providing users with helps that guide them through the analysis and development process. ASAT provides the user with a relational data base for storing, retrieving, and searching information. It includes a word processor and graphics editor for generating reports. A menu driven, user interface guides the users through the process of developing a front-end analysis, tasks listing, identifying critical tasks for individual and collective tasks, mission training plan, and lessons.

The automation of a job or portions thereof has been observed to have a significant impact on the ability of the incumbents to perform their jobs. An important but little mentioned impact is that the automation may very well indeed change the requirements (i.e., knowledge's skills, and abilities) needed to per-form the job/task. An example of this phenomenon is one we recently learned about in interviewing a trainer who had recently completed the automation of a job, "Tax Payer's Service Representative" for the Internal Revenue Service (IRS). The main responsibility of this representative is to answer questions that taxpayers might have about any tax codes. The task is to provide accurate information to the taxpayer. One of the goals of this automation was to make the job easier by off-loading a lot of the paper-based labor-intensive, time-consuming, repetitive tasks currently performed by job incumbents. Automation accomplished this by providing to the users information-managing tools custom-ized to enable them to perform their job with less effort. However, an unexpected outcome of this process of automation was that there was a significant change in the nature of the job itself, to the extent that the trainers had to provide retraining for the incumbents, in order for them to perform their job. In doing away with many of the paper-based tasks, they managed to remove many of the cues that triggered actions to be performed by the workers. In this way, the innovators not only made the process and procedures invisible to the job incumbents (personal communication S. Fisher, 1990) but also increased the complexity of the job itself. Norman (1988) referred to this as "the paradox of technology" and de-scribed it in the following way: "Technology offers the potential to make life easier and more enjoyable; each new technology provides increased benefits. At the same time, added complexities arise to increase our difficulty and frustration" (pp. 29–30).

An interesting problem for evaluators involved with the evaluation of automation is how to measure its effectiveness; often portions of the job have changed, and the knowledges, skills, and abilities required to perform the automated job have changed as well.

Changing Role of R&D in Computer Technology. Not only have we experienced a phenomenal growth in the number of and type of computers and other technology in our schools, businesses, and the military, but we have also witnessed changes in the focus of R&D educational technology and the ways that computer technology is used, and thus the types of evaluation questions that have been asked. For example, the DOE's focus of R&D technology, in the early 1970s, was on the use of television as an educational tool, the adaption of television to handicapped populations (e.g., closed-captioned television for the hearing impaired and deaf), application of telecommunications (e.g., electronic mail linking the Alaska school system) and satellite distribution of computer-based CAI instructional programing (Office of Technology Assessment, 1988). The focus of their effort was on the evaluation of the effectiveness of CAI and instructional television, as exemplified by the Education Testing Service (ETS) longitudinal evaluations of CAI effectiveness in the Los Angeles public schools and of children's "Sesame Street" television program. The main evaluation questions in these studies centered on outcome variables. Both studies were concerned with what was the effects of these innovations on children's achievement (outcome variables).

During the early 1970s NSF's funding was centered around the support of CAI (i.e., computer-based high school mathematics laboratory); computer languages (i.e., LOGO); and tools for aiding the development of CAI, such as; authoring languages and systems (PLATO IV and TICCIT). More recently, emphasis has shifted so that many of the advanced programs NSF has funded lately as demonstration projects are concerned through the use of intelligent tutors to teach calculus, algebra, geometry, prealgebra, and algorithmic problem solving (Office of Technology Assessment, 1988).

Here we note a shift from a focus of research questions involving the effectiveness of the technology as measured by achievement scores and other outcome variables to an emphasis on the impact of the technology on the educational or training environment (e.g., teachers' pedagogical strategies) and, more importantly, on students' cognitive organization, structure, and process.

A similar shift has occurred in the DOD, and several factors are responsible for this shift of emphasis. First, there has been a paradigm shift in psychological theory from a behavioristic to a cognitive view of learning. Second, there was a change in the military from emphasis on procedural training to a concern about complex problem solving. This shift in military training, in part, was brought about by the increasing demand for soldiers, airpersons, marines, and sailors with highly developed, complex problem-solving skills (Perez & Seidel, 1990). These skills are required to operate and maintain increasing sophisticated micro-electronic driven weapon systems. Next, there has been the emergence of advances in computer science and technology, which has provided the availability of powerful and relatively inexpensive personal computers, powerful graphic-oriented experimental workstations, the development of powerful and flexible object-oriented programing languages, and artificial intelligence technologies (such as expert systems). The latter have enabled the development of software programs that can emulate the problem solving-processes and skills of an expert within a specific domain (e.g., electronic troubleshooting and medical diagnosis). This shift in emphasis, as we see it, of R&D in the evaluation of the impact of computer technology is depicted in Fig. 7.1. The major shifts depicted of projects in Fig. 7.1 are away from traditional CAI to the use of computers as tools for learning and teaching, a major focus on the process of learning and teaching, rather than on outcomes per se. Finally, this paradigm shift represents an emphasis on the impact of the technology on the environment (e.g., teachers' pedagogical strategies), contextual variables (e.g., interactions between students and teachers) and the effects on students' cognitive process and structure, and generation of "deeper understanding" and development of mental models by the learners.

Evaluation of the Impact of Computer-Based Instruction

The promise of computer-based instruction has been to provide to each student high quality, inexpensive, individualized instruction. In military training and educational settings the promise has been partially met with computer-based

Current	Emerging
Emphasis on methods	Emphasis on content
Technology as teacher	Technology as tool
Engineering instruction	Engineering environments
Teaching for breadth	Teaching for depth ("less is more")
Emphasis on behavior	Emphasis on thinking
Teaching for performance	Teaching for understanding
Transmitting knowledge	Creating knowledge (constructivism)
Emphasis on teaching	Emphasis on learning
Media as teachers	Student as teachers
(teacher in a box)	(collaborative learning)

FIG. 7.1. Comparison of current and emerging uses of computer-mediated learning. Adapted from "Research and Evaluation Trends in the Uses of Computer-based Tools for Learning and Teaching" by H. Hunter (1991). *Proceedings of NECC 88.* Adapted by permission.

instruction demonstrating to be more cost-effective, requiring on the average 30% less time to train students for achieving equivalent or higher levels of performance than conventional training methods (Orlansky & String, 1977). Similar findings have been found in school settings by other researchers (J. A. Kulik, C. C. Kulik & Cohen, 1980; J. A. Kulik, Bangert, & Williams, 1983). Although positive effects of computer-based instruction have been demonstrated, that is, less time needed to train or educate than conventional methods, the proliferation of computers in schools, military installations, and the workplace has failed to materialize. One explanation of this is that the positive impact of computer-based instruction has not been attributed to computer technology but to the design of the instruction (Clark & Salomon, 1986). Regardless of the specific causal attribution, the introduction of computer technology to educational and training environments has *not* been fully realized in terms of the expected potential benefits, nor have these benefits been as widespread, or entirely successful as anticipated.

Kearsley, Hunter, and Seidel (1983) reviewed over 50 major CAI projects. Their review focused on the theoretical and practical significance of CAI to the field of education. In their review, the authors classified the projects as being one of eight categories; development of prototypes, conceptual demonstrations, major implementations and evaluations, disseminations, authoring language/systems, intelligent CAI, innovative environments, and new theory. They were able to identify nine major outcomes:

1. There is ample evidence that computers can make instruction more efficient or effective.
2. We know relatively little about how to individualize instruction.
3. We do not have a good understanding of the effects of instructional variables such as graphics, speech, motion, or humor.
4. A great deal has been learned about overcoming institutional and organizational inertia and resistance to change in the context of implementing CAI.
5. Significant progress has been made on the development of authoring tools and techniques for CAI.
6. Numerous mechanisms have been developed for the dissemination of CAI ideas and courseware.
7. CAI has spurred research throughout the entire field of instruction.
8. Federal funding has played a pivotal role in advancing CAI.
9. We have just scratched the surface of what can be accomplished with computers in education.

The authors go on further to conclude that CAI research has had substantial impact on education when assessed in total. We can expect an even greater impact in future decades as CAI technology becomes more powerful and accessible.

Meta-analytic Evaluations of CAI.   This section discusses the literature on the use of meta-analytic techniques to assess the effects of CAI on students learning. We think this literature is instructive in that it offers insight into the methodologies, and their limitations for evaluating the benefits of innovative computer technology.

Kulik and his colleagues (Kulik, 1992; J. A. Kulik et al., 1980; J. A. Kulik et al., 1983; Bangert-Drowns, J. A. Kulik, & C. C. Kulik, 1985) conducted several meta-analyses of the effects of CAI in a number of educational settings, including elementary, secondary school, and college. Their major evaluation question asked about the effects of CAI on student learning (achievement). Because so many research studies had been conducted to address this issue, they synthesized the results of these studies by using a meta-analysis technique. Meta-analysis is a statistical technique that combines the effects of independent research studies (see Glass, McGaw, & Smith, 1981, for a detailed discussion of this technique) by summarizing separate research results using a common statistical metric termed *effect size*. Effect size is estimated by using the sample means and standard deviations as reported in each study, or they can be calculated by using covariance-adjusted means or *t*-test statistics if available. Therefore, only those studies that report their standard deviations and means or provide statistics that can be used to derive them are included in the analysis. A frequently used method for calculating the effects size is the difference between outcome means of the experimental groups and the control groups, divided by the standard deviation of the control group. An alternative method used by Cohen (1977) is to divide the differences between the means by the pooled standard deviations of the experimental and control groups. Differences between the group means are stated as improvement or decrement in units of standard deviations. Researchers using this method when comparing the mean differences have used the following conventions to interpret the magnitude of the difference between means; less than or equal to .20 are interpreted as "small," whereas intermediate values between .20 and .80 are interpreted as "medium" and those greater or equal to .80 are considered to be "large" Cohen, 1977). Effect size is often transformed statistically into percentile scores for each group. Generally, when one conducts a meta-analysis, the researcher *a priori* establishes criteria for selecting studies to be included in the analysis. For example, J. A. Kulik and his colleagues only included those studies conducted in actual classrooms, where comparisons were made between groups receiving computer-based treatment and conventional treatments, and those studies that were free of serious methodological flaws.

The general finding from these studies is that computer-based education had positive effects on student achievement. The magnitude of the effects varies across elementary, secondary, and college samples. Studies with elementary school students found the average improvement was .47 standard deviations or a "medium" effect (Kulik, Kulik, & Bangert-Drowns, 1985). Studies with secondary school students found the average improvement was .40 standard deviations a "medium" effect (Kulik et al., 1985). In studies with college students the average improvement was the smallest, with .25 standard deviations (J. A. Kulik et al., 1980). J. A. Kulik, Kulik, and Schwalb (1986) conducted another meta-analytic study that included 24 controlled studies on adult education. Ten of these studies were concerned with military training. They found an improvement for CAI over conventional methods on the average of .42 standard deviations. In terms of percentile scores the results of these studies suggest that we should expect that CAI would raise the performance of the typical student from the 50th to the 66th percentile. Research on the effects of CAI has also found that CAI can reduce the time needed to train or educate. This conclusion has been supported by meta-analysis techniques (J. A. Kulik et al., 1980; C. C. Kulik et al., 1986) and conventional reviews of military training (Orlansky & String, 1979). In this latter study the authors reviewed the results of 19 studies and found that use of the computer saves trainees time in attaining the required minimum levels of knowledge and skills without the loss of student achievement (Orlansky, 1983). The median time saving in these studies was on the order of 30%. Kulik (1992), in reviewing the results of 10 separate meta-analyses concerned with answering questions about the effectiveness of computers in the classroom, concludes that

> all the meta-analysis that I have been able to locate show that adding computer-based instruction to a school program, on the average, improves the results of the program. But the meta-analysis differ somewhat on the size of the gains to be expected. We need to look more closely at the studies to determine which factors might cause variation in meta-analytic results. (p. 8)

Although these studies have provided policy-makers with evidence of the benefits that can be expected of CAI that can be used to guide decisions about the expenditures of resources, some researchers (Clark, 1985; Hagler & Knowlton, 1987; Shlechter, 1986; Salomon & Gardener, 1986) criticized these studies because they felt that comparing computer-based instruction with traditional treatments are confounded. That is, these researchers failed to control for variables such as content, teaching methods, or strategies, or the novelty of having a computer in the classroom.

The failure to match the content across treatment groups suggests that the quality of instructional materials may vary across the treatments, the amount of instruction and practice that each treatment group receives, and the level of difficulty of the materials. Seidel and Kopstein (1968, pp. 14–16) also noted that in the CAI condition the instructional model is explicit, whereas each "traditional" condition provides a sample of an implicit model of instruction.

An example of uncontrolled instructional quality is reported by Anderson,

Boyle, and Reiser, 1985). Where they report the results of a series of experiments where they compared the effectiveness of three treatment conditions in teaching, a computer programming language, *LISP*. The first experimental treatment condition consisted of students being taught *LISP* by an "Intelligent Tutor." The second treatment condition consisted of students being taught *LISP* by using paper-based materials that were based on cognitive model used in the first experimental conditions used to build the intelligent tutor. The third treatment was the traditional *LISP* course consisting of lectures and study exercises. The second treatment condition reported an advantage of the paper-based treatment over the regular classroom treatment by 10%. The only difference in the paper-based treatment materials was the organization of the materials were based on the cognitive model of *LISP* programing (Anderson, Farrell, & Sauers, 1984; Anderson et al., 1985). One could argue further that the objectives and content of the instruction between the control and the two experimental treatment were also different.

The previous criticisms have led some researchers to conclude that the positive impact of computer-based instruction may not be due to computer technology but to the design features of the instruction (Clark & Salomon, 1986). Clark and Solomon further argued that such outcome comparison studies with computers as the focus are indeed pointless because they fail to differentiate the causal effects of the technological media and other elements of instruction. Rather, they favor theory-based research or "holistic" descriptive studies, where the attributes and capacity of the computer to deliver varying instructional approaches are examined. However, these descriptive studies must also provide evidence for effects of the media on student outcomes. How else would we be able to identify what specific attribute(s) of the delivery media or the instructional design feature(s) is (are) responsible for the measured effect?

An even greater significant first step was presented concerning criteria matching stated purpose of use. Seidel (1980) argued that appropriate evaluation studies need to start from specifying the purpose for which the computer (or any other technology) is being used. J. A. Kulik (1994) performed a meta-analysis of 97 studies of CBI use in elementary and secondary schools. He divided CBI studies according to their use. He identified three uses of the computer in schools: tutor, tool, and tutee (after Taylor, 1980). The hypothesis he examined in this study supports that some approaches produce results that are better than average, whereas others will produce below-average results. Tutor studies are those where "the computer presents material, evaluates students reposes, determines what to present next, and keeps a record of the student progress" (J. A. Kulik, 1994). A distinction has to be made here between "tutors" that are traditional CAI systems and those computer-based systems that display some degree of intelligence (ICAI). The main difference between the two, other than the use of object-oriented computer languages and explicit models of cognitive process, is that the latter system diagnoses student responses on the "fly" and determines what is presented next. With traditional approaches the "what is presented next to the student" is preprogramed (see Park, Perez, & Seidel, 1987 for more detailed discussion of this point). The studies included in Kulik's analyses were of the CAI-type of tutor. The computer is defined as a tool when the computers used by a student for statistical analysis, calculation, word processing, spreadsheet, or word processing. The third type of use is when the computer serves as a "tutee" where the student uses programming languages, such as BASIC or Logo to give it direction. All three of these uses of the computer are discussed later.

In addition to computer usage Kulik used an approach of evaluating the degree of instructional innovations proposed by Slavin (1989). In this approach Slavin advocates that innovations can be defined with degrees of precision and that the precision with which an innovation is defined, will influence the success of its implementation and impact. He categorizes degree of precision in terms of three levels. Where at level 1, descriptions of the innovation are vague are not well defined, lack a conceptual basis, and are not well specified, where at level two would be a broadly defined notion like open-education or whole language instruction. Level 2 innovations are conceptually well defined, but their implementation varied depending on the interpretation of the practitioner. Examples of this level of innovation are computer instruction and cooperative learning. Level 3 innovations are, on the other hand, precisely defined including specific guidelines, or an explicit model, of the development of instructional materials and procedures for their use by teachers. Using these criteria Kulik performed three separate analyses. First, he examined all 97 studies together, which represented level 2. Then he analyzed the studies by grouping based on use, and finally he examined studies grouped together based on homogeneous subgroups of studies (see J. A. Kulik, 1994, for more detail). The average effect size in the first level of analysis was .32. His second level of analysis was positive, but he found a wide dispersion of effect sizes ranging from −.1 to .7 which suggests uncertainty in predicting the effects of CBI in a particular setting and that effect size is a function of specific computer use. The average effect size for those studies that had a well defined articulated model of instruction (Stanford-CCC program) Level 3, was .40, which means that for this program, gains of 1.4 years on a grade-equivalent, are likely with a year-long program. The results of Kulik's study suggest that use, purpose, context, and the maturity of an educational innovation are important attributes of CBI systems and must be included as a factors in any evaluation.

Kulik's study however, did not address Clark and Salomon's main criticism of CBI meta-analytic studies; that is, not being able to identify attributes of the delivery media or the instructional design feature(s) that is (are) responsible for the measured effect? One attempt to identify those attributes of the media that contribute to the measured effect is a study by Fletcher (1989). He conducted a meta-analytic study of the effectiveness of interactive videodisc (IVD). In this study he identified levels of interactivity of instruction and found that the more interactive features of IVD technology were used, the more effective the instruction. After an extensive review, Kozma (1985) argued that the separation of media and method is unnecessary. Because both are integrated by design, "the medium enables and constrains the method; the method draws on and instantiates the capabilities of the medium" (p. 34). However, it is precisely how well design features are implemented and the extent to which the attributes of the media are used that may determine the effectiveness of a CBI program.

Thus, it is important that evaluators of computer-based innovations consider and include the use, purpose, context, instructional design, and the maturity of an educational innovation in their evaluation design.

Evaluations of Intelligent Computer-Assisted Instruction.   Over the past decade there has been an increasing interest in computer-based instructional systems that display some degree of intelligence (ICAI). Sleeman and Brown (1982) suggested that the acronym ICAI be replaced by the acronym intelligent tutoring systems (ITS) to further distinguish instructional systems involving artificial intelligence from more traditional approaches. This chapter uses ITS. The ITS epitomizes the instructional ideal of individualized instruction. Although these systems have been in existence for over a decade, very few controlled comparative evaluations have been performed to determine their effectiveness. The lack of evaluation of ITSs has been due, in part, to the formative stage of development of these systems. Many of these ITSs were not completely developed systems, some had fully developed student diagnostic models, whereas others lacked mature, expert performance models. All of these components were deemed necessary before any systematic evaluation could be undertaken.

More recently, however, several ITSs have undergone systematic and controlled evaluation (Shute & Regian, 1990). The following tutors were evaluated by Shute and Regian: the LISP tutor (Anderson et al., 1984), Smithtown economics tutor (Shute & Glaser, 1991); Sherlock, a tutor for teaching electronic troubleshooting (Lesgold, Lajoie, Bunzo, & Eggan, 1991); and the Pascal tutor (Bonar, Cunningham, Beatty, & Weil, 1988). Shute and Regian (1990) summarized the results of these evaluations and examined the evidence for two claims made by ITS developers: (a) ITS would engender more effective learning in relation to traditional instructional methods; and (b) they would reduce the preexisting differences among learners on the posttest. The evaluation designs used in these studies generally compared the performance of the ITS group with some traditional method (e.g., lecture, reading materials, and on-the-job training) and a control group. The outcome measures included the time required to master material, time to solve problems or exercises, and performance on an outcome test of knowledge. The authors reported that evidence from these evaluations supported the first claim that ITS's are more efficient than traditional methods. Subjects working with the computer-based tutors (e.g., LISP tutor, Smithtown, Sherlock, Pascal tutor) acquired knowledge and skills in less time than the traditional instructional groups. However, no evidence was found to support the notion that computer-based tutoring would reduce the preexisting individual differences on posttest performance.

Although the results of these evaluations are not entirely positive, they are, encouraging. However, we still do not know what variables or design features of the tutors or their environments account for their effectiveness. The authors neglected to deal with two significant deficiencies in these evaluations: (a) the previously cited problems of comparing an explicit instructional model with a implicit instructional model, and (b) identifying the specific components contributing to the relative improvements in performance. This is of particular importance because a central theme in the work on ITSs is the need to adopt a unified view of working, learning, and innovating (Brown & Duguid, 1990). Lave and Wenger (1990) advocated that in order for training or educational programs to be effective and their trainees to be able to transfer information from one setting to another, these programs must include the complexities of practice and take into account the communities of practitioners. They must make provisions to incorporate the important dimensions of context often ignored in traditional views of transfer. The evaluation designs used to assess the effectiveness of ITSs have not yet begun to examine the social dynamics or important contextual variables.

Many of the evaluation studies discussed thus far have been designed to assess and demonstrate the impact of the computer primarily on student learning outcomes. Very few of these studies addressed important questions: What are the implications for thinking and learning in the classroom? What changes might we expect in the social context of the classroom? How might these innovations be used and how are they currently being used by users? Are they really effective? How do we evaluate their effectiveness? Effects of computer uses in education much like most other educational innovations are often difficult to implement and even more problematic to assess. The following sections will revisit some of these important issues in the context of a proposed framework for designing comprehensive solutions to these significant questions.

PART II: EVALUATION METHODS AND ISSUES

Formative Evaluation of Innovative Technology

Historically, formative evaluation has emerged as an important aspect of the evaluation of any educational technological innovation. Formative evaluation is distinguished from summative evaluation by the type of questions it seeks to answer, the stage of development of the product or program, and information it provides to decision makers. Scriven (1967) defined formative evaluation as the evaluation performed while the program is still "fluid" that is, while the program is considered to be in some stage of development; therefore a primary concern is the improvement of the program or product. Formative evaluation studies derive their own questions, methodologies, and interpretative structures (Hendricks, Montgomery, Mielke, & Fullilove, 1984). The purpose of formative evaluation of any innovation is to inform the stakeholders of the status of the development of a product or program. Because the products and uses of these innovations vary greatly, the design of the formative evaluation study must also vary. Some tools are designed to make the development of instruction more cost-effective than without these tools. Others may be designed to facilitate total educational reform. Still others are to enhance existing curricula. Each purpose demands its own formative evaluation design and criteria (Seidel, 1980).

Further, the capability of software and computers to be used in multiple ways and for multiple purposes in educational settings poses evaluators with new and interesting methodological problems. The primary objective of educational evaluation, including formative evaluation, is to determine whether or not the program is meeting the instructional objectives for which it was designed. In addition, an objective of formative evaluation research of an educational innovation must be to determine in the case of a software product whether it is "working." By working, we mean whether the code is executing as designed; we also would identify, if any, modifications that may improve its functioning. Hawkins and Kurland (1987) proposed a definition of working that we have modified. "Working" is defined in the following sense:

1. Do the members of the target audience understand the purpose of the courseware?
2. Can users manipulate its various parts?
3. Do the users enjoy interacting with it?
4. Are the directions clear?
5. Does the program run without errors (crashes)?
6. Do users understand and learn what the program is intended to do?

In addition, we would add the following to Hawkins and Kurland's list:

7. Is the program meeting the instructional objectives for which it was designed? (for educational software)
8. Does the code execute within tolerable parameters (e.g., acceptable response time, etc)?
9. Is the program/tool immune to changes that result from the various idiosyncratic ways that users might attempt to use it?
10. Is the process or topic that the product/tools are designed to aid or teach transparent to the user?
11. Are their any unattended side effects (e.g., additional skills are required than originally planned)?

Formative evaluation within this context has three primary objectives in improving the innovation: (a) to determine if the program is meeting the instructional objectives that it was designed for, (b) to determine if the product is usable and identify product improvement modifications, and (c) to determine whether the software tools are effective. These objectives sound as if they belong within a summative evaluation. However, we believe the development process must be iterative where at each phase of the development cycle the proponents of the innovation are assessing whether it is meeting the objective it was designed for and the impact it has on understanding and learning at the formative stage of software development. This iterative process would ensure that the end product is not only usable, but that it is effective in attaining its stated goals.

There are several strategies for conducting formative evaluation research on the usefulness of the software program or tool under development. A straightforward approach would be to select a sample of the target users and try out various parts of the program collecting informally data that would indicate whether specific parts of the program work. Additional strategies were suggested by Hawkins and Kurland (1987, p. 265):

1. Valuable information is gained by testing with students fragments of the existing programs, or programs that embody some of the ideas or interface features that are important aspects of the program.
2. Various components of the design of programs can be tested using other media with the target students, such as representing the screens and depicting the sequence on vugraphs on a projection screen, as in rapid prototyping.
3. As software prototype versions become available, they can be tried-out with a small, representative sample of target users.
4. Members of the target group (students, teachers, or administrators) can be interviewed about their needs, suggestions, and ideas.

This type of evaluation would appear to be both straightforward and absolutely essential for quality control. The implementation of such an evaluation, however, may very well be hampered by real world constraints of time and money.

Evaluation of the Impact of Technology Within Context

Seidel et al. (1978) and others (Fletcher, 1989; J. A. Kulik, 1994; Newman, 1989; Pea, 1988; Sheingold, 1983) argued that in order to understand and evaluate the effects of the introduction of computers to an educational environment (classroom), we need to deviate from the traditional experimental psychology approach of pre and post test procedures and instead look for alternative frameworks that place the emphasis of study on those processes of change that accompany the use of the technology. To understand and thus evaluate the interaction between humans and the use of computers, it is necessary to examine the critical role of cultural and individual interpretation of the technology and then examine the actual patterns of use. Seidel et al. (1978), Pea (1988), and Newman, (1989) called for a new methodology of evaluation based on three common observations: (a) There is a lack of definition of a uniform treatment; (b) Pre and Posttest evaluation designs are insufficient for interpreting any observed differences; (c) the length of time of treatment has been too short and presents problems of external validity to researchers in explaining complex human learning.

To elaborate, the introduction of the computer in itself does not constitute a uniform treatment. The most interesting, and for education the most challenging, characteristic of the computer is that it is not simply a device that can be used in a single prescribed way. It can and is used in multiple ways to achieve various end goals (Sheingold, 1983). Although the most common computer uses today within educational settings remain the drill-and-practice type and the teaching of programing (Becker, 1982, 1991), the uses of computers range from tutorials, simulations, and information managing tools to local are networks. With this perspective, a critical research question is what uses are being made of computers in the educational environment generally and of the classroom of interest specifically.

With the exception of drill-and-practice and tutorial uses of computer, most computer uses are flexible and under the control of the users. The use of a computer is therefore open to multiple interpretations and to many different approaches and uses in the classroom. Even if one use of the computer is the sole focus of the evaluation for example, programing the use may not be uniform. An illustration of such variety is the study of the effects of the use of LOGO programing on cognitive skills development in children.

In the project as reported by Hawkins (1987) the teachers' interpretation of the use of the LOGO varied across classrooms. One teacher perceived that the purpose of using LOGO was to teach computer literacy skills, whereas another felt that it was for the purpose of enrichment and only allowed the children to use the computer after they had completed all their other work. Only one teacher understood that the intended purpose was to teach LOGO so that the researchers could investigate it effects on children's cognitive skills, debugging, planning, and procedural reasoning. Therefore, it is not surprising, given these multiple interpretations of the research goal by the teachers, that the investigators did not find evidence supporting the hypothesis that learning to program in LOGO enhanced problem solving skills in children. More recently, J. A. Kulik (1994) using a meta-analytic approach found that when you control for use a positive effect size of .58 is observed. However, he found inconsistency in the results when he did not control for its use or the criterion measure used. One way to control for this variability in use and interpretation is to train teachers to do exactly the same thing in exactly the same way in their classrooms, so that we can assess the effects of a specific use, interpretation, and approach. Although in this case this approach would control for variability of use, as a general rule it would unfortunately eliminate one of the most important dimensions of the computer, namely, the ways in which teachers naturally and creatively interpret, work with, shape, and adapt the technology.

A second problem, noted earlier, of evaluating the effects of computer technology is with the often-used pre- and posttreatment evaluation methodology. There is an assumption that these effects can be measured in a reasonably short period of time generally a few weeks, rather than months, or a school year. Furthermore, these evaluations assume that the comparisons of the same measures between pre- and posttreatment will provide sufficient information for interpreting any observed differences.

Based on our experiences, we do not believe that the full effects of the computer are likely to be observed in a short time period. However, J. A. Kulik (1994), in examining the effects of duration of a study, suggested that for precollege investigations (elementary and secondary) short durations, lasting less then 4 weeks often produce stronger effects than do longer studies on the criterion measure. He explained that short studies may be better controlled, "and in our view measure more specific outcomes." However, longer studies are more ecologically valid but may produce more unattended benefits. The length of time of treatment has always presented problems of external validity to researchers (Campbell & Stanley, 1966). The difficulty of evaluation stems, in part, from the flexible nature of the technology, the process of interpreting it, of adapting to it, and adapting the technology to the purposes of the environment. Such a position is intrinsic to our evaluation model; we assert that the environmental variables relevant to classroom, school, school system, and community must be taken into account in order to achieve sustained effects of the computer innovation. Therefore, we suggest the expansion of the focus of the evaluation to include an assessment of the effects of the computer on the organization of the environment (classrooms, schools, school system, and community in which the school system is located). These issues give rise to several questions about the effects of the computer on the organization of the environment:

- What are computers and different kinds of software good for?
- How does the hardware fit into the organization of the classroom physical and social space?
- What can students learn from computer-based experiences?
- How should students be accountable for learning?
- How does technology and the learning it affords relate to traditional areas of the curriculum and to the traditional modes of learning?

We propose that the concept of the "environment" needs to be expanded to include effects related to the school, school system, and community in which the educational or training system is located. The addition of these levels in turn demands more in depth and flexible study of innovative effects. Some researchers (Newman, 1989; Pea, 1988; Sheingold et al., 1984) have argued for a more in depth approach to the evaluation of the impact of computer technology and have eloquently pointed out the limitations of conventional experiments (pre and post test) for studying the effects of the introduction of computer technology to the classrooms. There still remains a need for actually accomplishing a broad, systematic inquiry of the effects of the intervention. To identify what achievement variables were affected and what attributes of the technology are responsible for these effects, we must still be able to estimate whether the sample of classes included in the study was representative of other classrooms in the schools and school system. We need to consider whether measures used to study and chart the progress of the students in observations are reliable, valid, and generalizable. In short, we must be able to determine whether our observations are reliable, valid, and generalizable.

Stodolsky (1972) among others argued for a combined approach of qualitative and quantitative methodologies for evaluating effects of early education. For example, they cited the value of the addition of ethnographic data methods to collect qualitative data on context (i.e., multiple interpretations of the use and purpose of the technology) for use in the interpretation of quantitative data (e.g., test scores). Conducting more longitudinal studies is required to increase the length of the treatment and study and to ensure that the full effects of the computer technology are observed. To account for the variation in the implementation of a computer technology that by its nature encourages varied uses, purposes, and student activities, these observational methods could be used to document this variation and to develop hypothesis and test for the effects of each student's treatment. Each variation in the student's program could be treated as a separate treatment. Studies conducted by Perkins (1981) on differences in thinking of novices and experts and *Schonfeld's* (1985) studies of mathematical problem solving using a clinical interventionist, cognitive, subject-matter sensitive method are also examples of research that could be included in this evaluation design. When these inquiry methods are combined with experimental approaches, they would not only provide us with information on the educative effects of the introduction of computer technology in the classroom, evidence that children are benefiting in some objective demonstrable way, but also with an understanding of how these technologies impact on the instructional environment.

In sum, any model of evaluation of a technological innovation must provide for an examination of how the technology is being assimilated and accommodated into the educational environment. To assimilate the technology means to incorporate it into the existing environment without changing the environment.

To accommodate to the technology is defined as changing the environment to capitalize on the new things permitted by using the technology. This implies that attention has to be given by the evaluators to how the technology is being shaped by, and how it is facilitating changes in, the environment in which it is being introduced. Moreover, the model must acknowledge and chart the process of assimilation and accommodation of the technology with the expectation that the process is not likely to be uniform or consistent within and across environments. We propose and describe such a model later in the chapter.

Multiple Levels of Entry and Embedded Contexts of Learning

Berman and McLaughlin (1978) described the process of teacher adoption of new methods and materials. Martin (1987) described two concepts that have been particularly useful in the analysis and understanding of the adoption process in schools.

The first concept, *multiple level entry,* calls attention to the technology itself, and its ability to provide multiple levels of entry to the teachers and students who use it (Levin & Kareev, 1980). A technology that provides different pathways for achieving the same goals and activities is flexible to accommodate a range of individual approaches to problem solving. Thus, there may very well be more than one correct approach to solving a problem. Martin (1987) also noted the notion of context within which the teachers, students, and technology function. This is referred to as the *embedded context.* This means the broad and complex community environment in which students, teachers, and the technology function. This concept leads us away from simply looking at how the individual students or teachers perform particular tasks and toward examining those factors that mediate, interact, and support learning.

Multiple levels of entry suggests that by offering several paths into an activity, both experts and novices can use a software program for their respective purposes. In some cases this requires building levels of complexity into the software program. Thus, certain software programs can be used at different levels of complexity.

For example, once a user has become familiar with the parameters of a program, he can select more complex options that provide corresponding gains in program control or flexibility. From an evaluative perspective it also supports the point made earlier (see also Seidel, 1980) that users at different levels of sophistication will have varying objectives and goals; and the technological innovation must be judged accordingly.

The concept of embedded contexts of learning has importance for capturing the process of technology adoption in that learning tasks may be viewed as occurring within embedded contexts, each of which influences the performance of a task. Thus, for example, a teacher's use of a PC will be influenced by the routes through which machines are introduced into the classroom. How a teacher uses the computer will be highly influenced by whether the computer is introduced by another colleague who is a computer enthusiast, or mandated by the school district as part of an innovative technological program. These factors will be critical to the evaluator in interpreting the results of the evaluation, particularly as each implementation site's context and needs vary.

The need to discover and evaluate how well a technological innovation is working, how well it is designed to address the needs of the target users, and what data to collect that would lead to product improvement are all parts of a first step in an evaluation process. Many stakeholders may, in fact, be concerned with product improvement, changes in the instructional and work environment, improved teaching and training methods, and improved thinking and problem solving processes of students and workers attributed to the introduction of a computer technology. They still are concerned with the impact and the benefits of their investment.

NEED FOR A PROCESS MODEL OF EVALUATION

Earlier we reported on a study (Kearsley et al., 1983) that reviewed over 50 major CAI projects in a span of two decades focused on the theoretical and practical significance of CAI to the field of education. In this review, the authors classified the projects they examined as fitting within at least one of eight categories. These categories were development of prototypes, conceptual demonstrations, major implementations and evaluations, disseminations, authoring languages/systems, intelligent CAI, innovative environments, and new theory. These categories were based on several important dimensions: the objectives of the project, the uses of and purposes of the computer as a technological innovation, and the maturity of the project itself. The maturity of a project and the degree to which the planned activities have been implemented are important factors in what and how we are to examine and assess its effects. At the programatic level it is this documentation that will not only provide us with information as to what it is we are evaluating but ensure that we are not at risk of evaluating "a nonevent" (Charters & Jones, 1973): "At what point can one say, for example, that an innovation has been adopted? How does the investigator establish that a so-called new program is generically different from the one it replaced, other than the name by which it is called?" (p. 5).

In general, how is change in an educational environment to be measured? Thus, any model of evaluation designed to evaluate computer-based innovations must include some estimate of the maturity of the project and its component parts. One method of estimating the maturity of a project is to judge it within a framework of a model of technology transfer. In technology transfer we are examining the process by which a computer-based innovation or any educational innovation becomes institutionalized.

To summarize, this approach requires a paradigm shift with an emphasis on the impact of the technology: on the environment and context (e.g., new and varied types of interaction between students and teachers) and on the effects on students' cognitive process and structure.

We know relatively little about how to individualize instruction. We do not have a good understanding of the effects of instructional variables such as graphics, speech, motion, or humor. Researchers have failed to control for such variables as content, teaching methods or strategies, or the novelty of having a computer in the classroom. The failure to match the content across treatment groups means the level of difficulty of the materials is not equivalent. Kopstein and Seidel (1967), also noted that in the CAI condition the instructional model is explicit, whereas each traditional condition provides a sample of an implicit model of instruction. Added to the lack of a definition of a uniform treatment, pre- and posttest evaluation designs are insufficient for interpreting any observed differences, and the length of time of treatment has been too short and presents problems of external validity to researchers. What is required is a model that uses both quantitative and qualitative data collection methods, is longitudinal, and assesses level of implementation of the project.

Earlier in this chapter we reviewed a number of evaluation studies designed to assess the impact of computer-based technology in school, military, and industrial settings. These evaluation studies generally fit into three categories: (a) those studies that stressed the assessment of the impact of the computer innovation by examining the performance on achievement variables of students taught by the computer-based (experimental) treatment and those taught by a traditional method (control), (b) those studies that have focused on an assessment of the impact of computer-based innovation by examining the contextual process variables on the educational environments, (c) those studies that attempted to derive a cost-benefit gain as a result of the computer-based innovation.

We summarized the results of these evaluation studies, noted their limitations and advantages both at methodological and conceptual level, and suggested an evaluation approach that combined qualitative and quantitative methodologies for evaluating the impact of computer-based innovation programs. Additionally, we suggested that an important variable to be included is the maturity of the innovation.

The most serious limitation of studies that focus only on achievement or outcome data (e.g., J. A. Kulik et al., 1980; J. A. Kulik et al., 1983) as a measure of the impact of computer-based innovation is that they fail to examine process variables that would enable them to identify what variables account for the observed increase in performance. Thus, even though the meta-analytic studies can provide decision makers with an estimate of the size of the effect of the technology, they limit our understanding of how the technology works and do not contribute to the formulation of theory.

Those studies that have been focused on an assessment of the impact of computer-based innovation by examining the contextual and process variables (e.g., Newman, 1989; Pea, 1988; and Sheingold, 1983) have provided rich descriptions of how the computer is used, how it has influenced the cognitive processes of students, and how it has changed the educational environment. These studies could further contribute both to our understanding of attributes of the computer and to theory building. However, these studies do not provide an estimate of the effect of technology on important educational goals or objectives. Thus, we know much about how these technologies are changing the way the students learn and teachers teach; but we do not know whether these changes are generalizable or produce significant effects. The missing bridge is the link between the objectives to be served and the measurements obtained. As noted, each project purpose demands its own evaluation design and criteria (Seidel, 1980). If cost-effectiveness is the goal, then alternative methods and their cost to achieve the same objectives must be measured. If curricular enhancement is the reason for the computer, then the new objectives, activities, and processes made available to learner and instructor need to be the focus for evaluation. We now turn to an approach that takes into account the numerous and diverse variables already discussed and yields a framework for meaningful synthesis: the surrounding environment.

A PROPOSED MODEL

We propose at this point a model for evaluating technological innovation in training or educational environments. Our model considers purpose, process, outcome measures, and the maturity of the innovation. In our model of evaluation of technology their are three major stages—adoption, implementation, and institutionalization—and two major processes: assimilation and accommodation of the innovation. These stages and the processes occur within a context of multiple levels. Taken together, they make up the dimensions of our evaluation model. Our primary goal is technology transfer, which will ensure that the innovation becomes institutionalized within the educational or training setting. Taking the computer as our example, the first stage is to adopt the use of the computer. What do we mean by that? Adoption occurs with the initial use of new technology within an administrative unit; for example in a course within a department, whether it be at a school or a unit, the single use of the new technology is adoption. Often the adoption stage is confused with the institutionalization stage. That is, it is taken for granted that it has become accepted as an integral part of the training establishment. The field is rife with examples of failures in this respect. For illustration we choose just one failure of an innovation to become institutionalized because it is a dramatic one. The automated instructional system (AIS) in the airforce was initially started as a system to be institutionalized once demonstrated to be successful. It had a very successful initial adoption within a few targeted courses at Lowery Air Force Base. However, since the initial adoption, there were many administrative, organizational, and individual reasons—political in many cases to be sure—that prevented that system ever from reaching its full potential as an institutional part of the airforce training package.

The next stage to be considered in the development of the use of the computer is implementation. In this stage, the innovation is used beyond the initial adoption, such as in a typical department or course; it is used across the board in an entire school system and within a local organizational structure. An example of an innovation that it has been implemented in a military setting is the automated resources scheduler at Fort Rucker used in army helicopter training.

However, as noted in Fig. 7.2, the goal is to institutionalize the innovation; and that stage only happens when the innovation itself is an integral part of the entire educational or military training system.

In considering the model for technology transfer and reaching the goal of institutionalizing the computer innovation in education and training, we should all be aware that any innovation goes through two processes; assimilation and accommodation. Assimilation is characterized by the incorporation or absorption of the innovation into the existing environment (e.g., user applying the computer in much the same way they used pencils and paper in the past). By this we mean the innovation helps us to do something we ordinarily do, but it helps us to do it better, or faster, or a combination thereof. For example, if we were performing

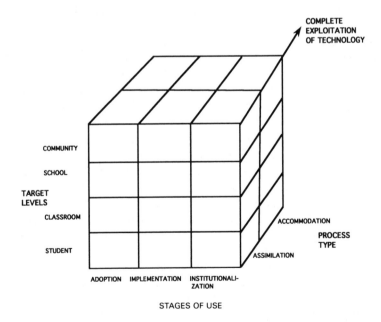

FIG. 7.2. A Model for technology evaluation.

record keeping by hand, we would assimilate the use of the computer into doing the same record keeping, albeit more efficiently. In terms of delivery of instruction, we would use the computer the way we have utilized blackboards, overheads, slide projectors, and other media. Therefore, it would be relatively easy to convince people to use the computer to give remedial drill-and-practice instruction while retaining the same basic content of instruction.

Accommodation is characterized by the adaptation or adjustment of the environment to the technological innovation. An example of this process is reported by Newman (1990) in which the researchers introduced a local area network (LAN) into an elementary school. The goal of this project was to increase the frequency of collaborative work among the students. They not only found, that the frequency of collaborative work increases among the students, but that teachers involved in the study adapted the use of the LAN to increase their collaborative efforts with other teachers.

The real value of an innovation of advanced technologies occurs when we reach the point that we can accommodate to the value added, to the uniqueness, to the flexibility that the technological innovation provides us. Very rarely has that occurred, at all, with respect to computers in training. The potential for this, of course, occurs with being able to perform tasks in training that we would otherwise not be able to perform. The use of simulations, not just simulators, that faithfully represent not only the same tasks as the operational equipment but the use of computer-based simulations with greater degrees of flexibility than the actual equipment. This enables us to go beyond the capability of the actual equipment with the possibility of doing jobs or creating conditions never conceived of before using the computer. For example, during a site visit at the University of California at Irvine in 1974, one of us witnessed a demonstration of the use of computer generated dialogues in dealing with the prediction of trajectories of bullets or shells as they reached their target. There were certain variables, including speed, length of the gun barrel, varying velocities, and so on, that could be manipulated by the users of this particular exercise. The results of these manipulations of the variables was a counter-intuitive outcome that included the fact the bullet or shell would never reach, in certain instances, the end of the barrel of the gun. A situation that cannot be readily performed using the actual equipment. This is trivial, perhaps, but at the same time it was counter-intuitive and a surprising finding for some highly experienced physicists. There are other examples of accommodating to the computer, such as the fact that we currently have administrative requirements of so many hours per day, per class, or per course, to be accomplished in order for a trainee or a student to complete his instruction. The point is, the computer knows no such administrative limits. It only operates on a competency-based model of learning. It can be the vehicle for changing the administrative model (now group based) to match a new computerized instructional model, based on progress of the individual student assignment within a course and therefore, individualized course completion, assignment to further training/education, and ultimately individualized graduation.

We will reach our full accommodation to the use of the technological innovation when we allow for appropriate competency-based models of individualized learning coupled with individualized selection and assignment of trainees. This model clearly requires changes in the organizational structure of the training and education establishments if institutionalization of the exploitable capabilities of the computer is to occur.

As a result of these observations, we feel that in order for institutionalization to take place, a proper model for exploiting the technological innovation therefore requires a linking of the dimensions of research, management of instruction and training, and the operational education/training itself.

Within the army, implementation of a pilot program was accomplished by the establishment of the Training Technologies Field Activities (TTFA) program.[1] This required the Army Research Institute, the R&D arm of the army, to work closely with the TTFA office at the Training and Doctrine Command Headquarters (TRADOC), as well as with the pilot TRADOC schools which were the targeted training sites for this new program. Projects were chosen that focused on different psychological skills (perceptual motor, complex problem solving, and procedural skills) along with varied requirements for types of trainees (commissioned officers, noncommissioned officers, and enlisted personnel). These occurred within a spectrum of different costs, which varied for active versus reserve components of the army. The products developed by this approach were unique to the schools, such as a procurement for improving a particular targeted course at the school; and some technologies were to be generalizable across all schools, such as a technique for doing training data management (broadly speaking computer managed instruction). Another was the development of the technology transfer model to apply to all army schools once these four pilots efforts were underway. Although TTFA enjoyed a modicum of success, unfortunately, because of budget cuts, much of the infrastructure to support the TTFA was removed and the program was terminated.

Such an approach involves the entire community that surrounds the attempt at studying a particular technological innovation. The analogy is clear between education and this military example in which we have been studying the relationship between the user community and its wider operational environment plus the institutional environment, for example, their school, the particular classroom environment involving the targeted course, and the instructor. Given that all of

---

[1]This program was curtailed based on cut-back of funds for both training and R&D within the army. However, some of the accomplishments still exist, although the broader-based institutionalized goal was not achieved.

the previous elements have been appropriately involved in and accept the innovation, the last event, of course, is the training of the students with the particular application of the computer-based innovation.

Current research and evaluation in the use of the computers as tools should seek a deeper understanding of the context in which these tools are used rather than simply asking the question: Does the computer make a difference in learning, an educational outcome, a specific task, or topic?

The reader might think of the context of evaluation having multiple levels like a target with concentric circles. The bulls eye is student use. Each surrounding ring enlarges the area of concern to include another important level of influence on whether or not the innovation gets used as intended. From curricular/instructor environment, through school, system, and surrounding influential community, all levels comprise the necessary and sufficient conditions to permit appropriate computer use and assessment thereof. Moreover, we must proceed from the outside levels into the center of the circle to insure that we will avoid the risk noted earlier of trying to measure student achievement when the other levels prevent this (nonevent measurement problem).

At this point, it seems appropriate to present a synthesis of the nine principal functions of the model. We propose this model as a meaningful three-dimensional representation of the technology evaluation in any training/educational environment.

The target levels comprise the interested parties that can make or break the success of an innovation, so called stakeholders. From the training or educational perspective, most often there is a multitude of users but generally the student/trainee is the ultimate user. To determine whether or not the technology is beneficial, other groups must be committed toward allowing the innovation to be used as intended. The instructor has to fit the technology into a curriculum. Will it be the same curriculum or a new one? Will the instructor be trained to use the technology to do something different or teach the same topics more effectively? At the next level, the question is how committed is the principal, the administrator (in the army, the school commandant) and/or the faculty to using the new technology? All these stakeholders influence the allocations of resources space, time, and other resources required to use the technology properly.

The outer level of interested parties is the user community within which the school (s) is (are) located. Again, support for the computer use, budget allocations for new hardware or maintenance all must be satisfied at a systemwide level. In the educational community this means policy-makers, politicians, teachers unions and associations, PTAs, business leaders, and local citizenry. In the army this means the Training and Doctrine Command (TRADOC) responsible for setting training policy, the commandants of the various schools, generals who head major commands, the secretary of the army, and at the local level of department staff responsible for training operations.

According to the model, typically it would be easiest to convince an administrative group, budget comptrollers, or instructors to use a technology initially to accomplish a familiar or similar usage in a single course (assimilation usage). Changing curricula markedly, however, whether with LOGO as Papert (1980) advocated (e.g., Mindstorms) with microworlds (White & Frederiksen, 1986); or with large scale educational reforms by others, will take much longer to accomplish. Such wholesale accommodation to the technology would require all levels of stakeholders to accept the value-added potential of institutionalizing such use of the computer. It is not surprising, therefore, that after four decades of computer projects in training and education there are no instances of complete exploitation of computer technology.

How do we enhance the likelihood of successful, novel computer use, institutionalized in a training/education system? The solution would seem to be first convince the decision-making bodies to take a small adoption step, proceed to schoolwide implementation of a familiar use and then institutionalize this (assimilation). Having been successful, along the one face of the model, it would seem that accommodation to the technology's novel features could be made easier. Of course, another alternative is to establish experimental schools where all types of use could be accomplished and the stages telescoped. Variations on these two extreme themes are also possible. But the key to any attempt at technological innovation in the education or training communities is to involve all levels of stakeholders up front and keep them involved throughout the proposed processes of assimilation and accommodation.

Evaluation measures must be devised to track amount of involvement by all target levels. This means that process measures as well as student/trainee achievement measures need to be considered. For example, in the illustrations cited earlier, TTFAs in the army and the District of Columbia Secondary Schools Project for Adopting Computer-Aided Education (DCSSPACE) projects are examples of how this model has evolved over the years. The DCSSPACE project sponsored by NSF, involved the introduction of computers and previously developed computer curricular materials into the secondary schools of the District of Columbia and then studying their adoption. Because of the uniqueness of each school system, the specific evaluation measures will best be developed on a case-by-case basis.

What strategies can be said to facilitate the institutionalizing process across the multiple levels? The elements of a strategy reside in the "Three C's": control, communication and commitment. At each level, students, classroom teachers, school administrators, parents, and community leaders must perceive that they each have a share of *control* over the degree and type of use of the technological innovation. To aid in achieving this state, continual and open, reciprocal communication across levels during the stages of adoption must be followed. Various techniques exist to accomplish this: A successful one is Transactional Evaluation (Wagner & Seidel, 1978). It basically involves repeated attitudinal checks at regular intervals with follow-up strategies to eliminate problems of misunderstanding (polarization, ambiguity, etc.). When control and communication are in place, then and only then can commitment to adopt the new technology follow. This means motivation by the student to learn, commitment by the teacher to modify the curricula in order to accommodate to the technology, desire by the school administration to aid this change, and lastly, resources provided by the community boards to enable timely institutionalization. The approach to the evaluation of technological innovations in training/educational environments must be directed at the aforementioned concerns stated here. It must do so to answer significant education questions: What are the implications for thinking and learning in the classroom? What changes might we expect in the social context of the classroom? How might these innovations best be used and how are they currently being used by users? And, most important, are they really effective? How do we evaluate their effectiveness? It is these issues, among others, that are the focus of this proposal. To address them adequately requires any evaluator to take into account all the important functions to be performed in the use and adoption of any innovative technology. To focus solely on technical, costs, or outcome measures would ignore the most critical factors in the adoption process and the assessment of the impact of the educational innovation.

This general approach has evolved from observing various aspects of the training/educational settings and from a continual review of the literature. It is anticipated that this approach will satisfy the need for a systematic yet flexible method of gathering, analyzing, and presenting information bearing on a wide range of complex and interrelated factors. Our proposed model examines such variables as attitudes, values, roles, interpersonal relationships, and the social context, along with "objectives" as well as measures of student achievement, attendance, computer usage, and cost factors. We hope the model will provide useful suggestions for a new generation of education/training technology evaluators.

SUMMARY

In this chapter our principal goal has been to present a comprehensive model for evaluating technological innovations in training/educational environments. In order to do so, we first set the stage by providing examples of historical and more recent attempts and their deficiencies. Next, we covered the requirements for a more complete model of assessment. We then presented a multi-dimensional model that considers the levels of surrounding environments within which the innovation occurs. We highlighted the requirement for measuring necessary processes as well as products of using these innovations. Lastly, we described the stages of incorporating the innovations into the training/education systems, which stages ultimately can lead to complete exploitation of the innovation. Examples were provided of these dimensions and suggestions were given for implementing the model.

DISCLAIMER

The opinions expressed herein are those of the authors and in no way represent any official position of any agency of the U.S. Government.

REFERENCES

Anderson, J. R., Boyle, C., & Reiser, B. (1985). Intelligent tutoring systems. *Science, 228,* 456–462.

Anderson, J. R., Farrell, R., & Sauers, R. (1984). Learning to program in LISP. *Cognitive Science, 8,* 87–129.

Baker, E. (1990). Technology assessment: Policy and methodological issues. In H. L. Burns, J. Parlett, & C. Luckhardt (Eds.), *Intelligent tutoring systems: Evolutions in design.* Hillsdale, NJ: Lawrence Erlbaum Associates.

Barr, A., & Feigenbaum, E. A. (Eds.). (1982). *The handbook of artificial intelligence* (Vol. 2). Los Altos CA: William Kaufmann.

Bangert-Drowns, R. L., Kulik, J. A., & Kulik, C. C. (1985). Effectiveness of computer-based education in secondary schools. *Journal of Computer-Based Instruction, 12*(3), 59–68.

Becker, H. J. (1982). *Microcomputers in the classroom: Dreams and realities* (Rep. No. 319). Baltimore, MD: Johns Hopkins University, Center for Social Organization of Schools.

Becker, H. J. (1991). How computers are used in United States schools: Basic data from the I.E.A. computers in education survey. *Journal of Educational Computing Research, 7* (4), 385–406.

Berman, P., & McLaughlin, M. (1978). *Federal programs supporting educational change, Vol. 3. Implementing and sustaining innovations* (Rep. No. R-1589/8-HEW). Santa Monica, CA: The Rand Corporation.

Bonar, J., Cunningham, R., Beatty, P., & Weil, W. (1988). *Bridge: Intelligent tutoring system with intermediate representations* (Tech. Rep. No. AIP-21). Pittsburgh: Learning Research & Development Center, University of Pittsburgh.

Brown, J. S., & Duguid, P. (1990). Organizational learning and communities-of-practice: Toward a unified view of working, learning, and innovation *Organizational learning, 2*(1), 40–57.

Burton, R. R., & Brown, J. S. (1982). An investigation of computer coaching for informal learning activities. In D. Sleeman & J. S. Brown (Eds.), *Intelligent tutoring systems* (pp. 79–87). London, England: Academic Press.

Boroughs, D. L., Black, R. F., Ferguson, G., & Pasternak, D. (1990). Desktop dilemma. *U.S. News & World Report,* 24 December, pp. 46–48.

Campbell, D. T., & Stanley, J. C. (1966). *Experimental and quasi-experimental designs for research.* Chicago: Rand McNally.

Carey, S. (1986). Cognitive science and science education. *American Psychologist, 41*(10), 1123–1130.

Charters, W. W., & Jones, J. E. (1973). On the risk of appraising non-events in program evaluation. *Educational Researcher, 2*(11), 5–7.

Clark, R. E. (1985). Confounding in educational computing research. *Journal of Educational Computing Research, 1*(2), 137–148.

Clark, R. E., & Salomon, G. (1986). Media in training. In M. C. Wittrock (Ed.), *Handbook of research on teaching* (3rd ed., pp. 464–478). New York: Macmillan.

Cohen, J. (1977). *Statistical power analysis for the behavioral sciences* (rev. ed.). New York: Academic Press.

DeSanctis, G., & Gallupe, B. (1987). A foundation for the study of group decision support systems. *Management Science, 33,* 589–609.

Fletcher, J. D., & Rockway, M. R. (1986). Computer-based training in the military. In J. A. Ellis (Ed.), *Military contributions to instructional technology* (pp. 171–222). New York: Praeger scientific press.

Fletcher, J. D. (1989). The effectiveness of interactive videodisc instruction. *Machine-Mediated Learning, 3,* 361–385.

Fletcher, J. D., Hawley, D. E., & Piele, P. K. (1990). Costs, effects, and utility of microcomputer assisted instruction in the classroom. *American Educational Research Journal, 33*(4), 783–806.

Glass, G. V., McGaw, B., & Smith, M. L. (1981). *Meta-analysis in social research.* Beverly Hills, CA: Sage Publications.

Halff, H. M., Holland, J. D., & Hutchins, E. L. (1986). Cognitive science and military training. *American Psychologist, 41*(10), 1131–1139.

Hagler, P., & Knowlton, J. (1987). Invalid implicit assumption in CBI comparison research. *Journal of Computer-Based Instruction, 14*(3), 84–88.

Hawkins, J. (1987). The interpretation of logo practice. In R. D. Pea & K. Sheingold (Eds.), *Mirrors of minds: Patterns of experience in educational computing* (pp. 3–34). Norwood, NJ: Ablex.

Hawkins, J., & Kurland, D. M. (1987). Informing the design of software through context-based research. In R. D. Pea & K. Sheingold (Eds.), *Mirrors of minds: Patterns of experience in educational computing* (pp. 258–273). Norwood, NJ: Ablex.

Hawkins, J., Sheingold, K., & Berger, C. (1982). Microcomputers in schools: Impact on the social life of elementary classrooms. *Journal of Applied Developmental Psychology, 3,* 361–373.

Hendricks, L., Montgomery, T., & Fullilove, R. (1984). Educational achievement and locus of control among Black adolescent fathers. *Journal of Negro Education, 53*(2), 182–188.

Hiltz, S. R., Johnson, K., & Turnoff, M. (1986). Experiments in group decision making: Communication process and outcome in face-to-face versus computerized conferences. *Human Communication Research, 13,* 225–252.

Hunter, B. (1988). Research and evaluation trends in the uses of computer-based tools for learning and teaching. *Proceedings of NECC 88,* Dallas, pp. 82–93.

Hunter, H. (1991). *Emerging training technology trends.* Unpublished paper presented at the Intergovernmental Agency Group monthly meeting.

Kearsley, G., Hunter, B., & Seidel, R. J. (1983). *Two decades of CBI research: What have we learned?* (HumRRO Professional Paper No.3-83). Human Resources Research Organization, Alexandria, VA.

Keisler, S. Siegel, J., & McGuire, T. W. (1984). Social psychological aspects of computer-mediated communication. *American Psychologist, 39,* 1123–34.

Kopstein, F. F., & Seidel, R. J. (1967). "Comment on Shurdak's An approach to the use of computers in the instructional process and an evaluation. *American Educational Research Journal, 4,* 413–416.

Kozma, R. B. (1991). Learning with media, review of educational research, *Review of Educational Research, 61*(3) 179–211.

Kulik, C. C., Kulik, J. A., & Schwalb, B. J. (1982). Programmed instruction in secondary education: A meta-analysis of evaluation findings. *Journal of Educational Research, 75*(3), 133–138.

Kulik, J. A. (1994). Meta-analytic studies of findings on computer-based instruction. In E. L. Baker & H. F. O'Neil, Jr. (Eds.), *Technology assessment in education and training.* Hillsdale, NJ: Lawrence Erlbaum Associates.

Kulik, J. A., Bangert, R. L., & Williams, G. W. (1983). Effects of computer-based teaching on secondary school students. *Journal of Educational Psychology, 75*(1), 19–26.

Kulik, J. A., Kulik, C. C., & Bangert-Drowns, R. L. (1985). Effectiveness of computer-based education in elementary schools. *Computers in human behavior, 1,* 59–74.

Kulik, J. A., Kulik, C. C., & Schwalb, B. J. (1986). The effectiveness of computer-based adult education; A meta-analysis. *Journal of educational research, 75,* 133–138.

Kulik, J. A., Kulik, C. C., & Cohen, P. A. (1980). Effectiveness of computer-based college teaching: A meta-analysis of findings. *Review of Educational Research, 50*(4), 525–544.

Kurland, D. M., Catherine C. A., Mawby, R., & Pea, R. D. (1987). Mapping the cognitive demands of learning to program. In R. D. Pea & K. Sheingold (Eds.), *Mirrors of minds: Patterns of experience in educational computing* (pp. 103–127). Norwood, NJ: Ablex.

Lave, J., & Wenger, E. (1990). Situated learning: Legitimate peripheral participation (IRL Rep.No. 90-0013). Palo Alto, CA: Institute for Research on Learning.

Lesgold, A., Lajoie, S. P., Bunzo, M., & Eggan, G. (1991). A coached practice environment for an electronics troubleshooting job. In J. Larkin, R. Chabay, & C. Sheftic (Eds.), *Computer-assisted instruction and intelligent tutoring systems: Establishing communication and collaboration* (pp. 223–256). Hillsdale, NJ: Lawrence Erlbaum Associates.

Levin, J. A., & Kareev, Y. (1980). *Personal computers and education. The challenge to schools* (CHIP Report No. 98). La Jolla CA: University of California, San Diego, Center for Human Information Processing.

Lewis, M. W., McArthur, D., Stasz, C. & Zmuidzinas, M. (1990). *Discovery-based tutoring in mathematics* (Working Notes). AAAI Spring Symposium Series, Stanford University, CA.

Liao, Y. K. (1992). Effects of computer-assisted instruction on cognitive outcomes: A meta-analysis. *Journal of Research on Computing Education, 24*(3), 367–80.

McGuire, T. W. Keisler, S., & Siegal, J. (1987). Group and computer-mediated discussion effects in risk decision making. *Journal of Personality and Social Psychology, 52,* 917–30.

Mielke, P. (1982). *Military Manpower Training Report for FY 1987, Vol. 4: Force Readiness Report.* Department of Defense.

Newman, D. (1987). Functional environments for microcomputers in education. In R. D. Pea & K. Sheingold (Eds.), *Mirrors of minds: Patterns of experience in educational computing* (pp. 57–66). Norwood, NJ: Ablex.

Newman, D. (1990). Opportunities for research on the organizational impact of school computers. *Educational Researcher, 19*(3), 8–13.

Norman, D. A. (1988). *The psychology of everyday things.* New York: Basic Books.

Office of Technology Assessment (1988). *Power on: New tools for teaching and learning.*

O'Neil, H. F. Jr., & Baker, E. L. (1991). Issues in intelligent computer-assisted instruction: evaluation and measurement. In T. B. Gutkin & S. L. Wise (Eds.), *The computer and the decision-making process* (pp. 199–224). Hillsdale, NJ: Lawrence Erlbaum Associates.

Orlansky, J., & String, J. (1979). *Cost effectiveness of computer-based instruction in military training* (IDA Paper P-1375) Institute for Defense Analysis.

Orlansky, J., & String, J. (1981). Computer-based instruction in military training. *Defense Management Journal, 18*(2), 46–54.

Papert, S. (1980). *Mindstorms.* New York: Basic Books.

Park, O. K., Perez, R. S., & Seidel, R. J. (1987). Intelligent CAI: Old wine in new bottles or a new vintage? In G. P. Kearsley (Ed.), *Artificial intelligence and instruction applications and methods.* Reading, MA: Addison-Wesley.

Pea, R. D. (1985). Beyond amplification: Using the computer to reorganize mental functioning. *Educational psychologist, 20,*(4), 167–182.

Pea, R. D. (1987). Integrating human and computer intelligence. In R. D. Pea & K. Sheingold (Eds.), *Mirrors of minds: Patterns of experience in educational computing* (pp. 128–146). Norwood, NJ: Ablex.

Pea, R. D. (1988). *Distributed intelligence in learning and reasoning processes.* Paper presented at the meeting of the Cognitive Science Society, Montreal.

Pea, R. D., & Kurland, D. M. (1987). On the cognitive effects of learning computer programing. In R. D. Pea & K. Sheingold (Eds.), *Mirrors of minds: Patterns of experience in educational computing* (pp. 147–177). Norwood, NJ: Ablex.

Pea, R. D., & Sheingold, K. (Eds.). (1987). *Mirrors of minds: Patterns of experience in educational computing.* Norwood, NJ: Ablex.

Perez, R. S., Gregory, M., & Minionis, D. (1992). Tools and decision aids for training development in the U.S. Army. Special Issue of *Instructional Science.*

Perez, R. S., & Seidel, R. J. (1990). Using artificial intelligence: Computer-based tools for instructional development. *Educational Technology, 30*(3)51–59.

Perez, R. S., & Seidel, R. J. (1986). Cognitive theory of technical training. In T. Sticht, F. Chang, & S. Wood (Eds.), *Advances in reading/language research* (pp. 139–166). Greenwich Ct. JAI.

Perkins, D. (1981). *The mind's best work.* Cambridge, MA: Harvard University Press.

Salomon, G., & Gardener, H. (1986). The computer as educator: Lessons from television research. *Educational Researcher, 15*(10), 13–19.

Schoenfeld, A. H. (1985). *Mathematical problem-solving.* Orlando, FL: Academic Press.

Scriven, M. (1967). The methodology of evaluation. In R. E. Stake (Eds.), *Perspectives on curriculum evaluation.* AERA Monograph series on Curriculum Evaluation, No. 1. Chicago: Rand McNally.

Schlechter, T. M. (1986). *An examination of the research evidence for computer-based instruction in military training.* U.S. Army Research Institute for the Behavioral and Social Sciences, ARI Field Unit, Fort Knox, Kentucky.

Seidel, R. J. (1974, January). Educational technologies: Higher education. In S. Harrison & L. Stolurow (Eds.), *Proceedings of Productivity in Higher Education Symposium,* pp. 157–158.

Seidel, R. J. (1978). *D.C. SSPACE Project,* Final Report. (Grant. no. EEP74–19456). National Science Foundation.

Seidel, R. J. (1980). It's 1980: Do you know where your computer is? *Phi Delta Kappan,* 481–485.

Seidel, R. J., & Kopstein, F. F. (1968). *A general systems approach to the development and maintenance of optimal learning conditions.* HumRRO, professional paper 1–68. Washington D.C.

Seidel, R. J., & Rubin, M. L. (1977). Introduction, summary and implications. In R. J. Seidel & M. L. Rubin (Eds.), *Computers and communication: Implications for education* (pp. xvii–xx) New York: Academic Press.

Seidel, R. J., & Wagner, H. (1979). A cost effectiveness specification. In H. F. O'Neil (Ed.), *Procedures of instructional systems development* (pp. 233–250). New York: Academic press.

Sheingold, K., Hawkins, J., & Char, C. (1984). I'm the thinkist, you're the typist: The interaction of technology and the social life of classrooms. *Journal of Social Issues, 40*(3), 49–61.

Shute, V. J., & Glaser, R. (1990). A large scale evaluation of an intelligent discovery world: Smithtown. *Interactive Learning Environment, 1,* 51–77.

Shute, V. J., & Regian, J. W. (1990, May). *Rose garden promises of intelligent tutoring systems: Blossom or thorn?* Paper presented at the Space Operations, Applications and Research (SOAR) Symposium, Albuquerque, NM.

Slavin, R. E. (1989). Pet and pedulum: Fadism in education and how to stop it, *Phi Delta Kappan, 70* (10), 752–58.

Sleeman, D., & Brown, J. S. (1982). *Intelligent tutoring systems.* New York: Academic Press.

Strassman, P. A. (1990). *The business value of computers.* New Canaan, CT: The Information Economic Press.

Stodlsky, S. (1972). Defining treatment and outcome in early childhood education. In H. J. Walberg & A. T. Kopan (Eds.), *Rethinking Urban Education.* San Francisco: Jossey-Bass.

Taylor, R. P. (Ed.). (1980). *The computer in the school: Tutor, tool, tutee.* New York: The Teachers College Press.

Tobias, S. (1984). Computers in the classroom. *Contemporary education review, 3*(2) 387–390.

Wagner, H., & Seidel, R. J. (1978). Program evaluation. In H. O'Neil (Ed.), *Learning strategies.* New York: Academic Press.

White, B. Y., & Frederiksen, J. R. (1986). Intelligent tutoring system based upon qualitative model evolutions. *Proceedings of AAI-86,* Los Altos CA: Morgan Kaufman, pp. 313–319.

8 Visualization Tools for Model-Based Inquiry

Wallace Feurzeig
BBN Systems and Technologies, Cambridge, MA

Computer simulation is rapidly becoming a standard part of the methodology of science research. It bridges the two established science inquiry paradigms, experiment and theory, informing and enhancing both. The outputs generated in science simulations are routinely expressed visually as computer animations. Animated displays of the modeled behaviors often give valuable insights about underlying processes. This chapter describes tools for visualization of computer simulations in science education and research and presents an informal assessment of the technological approaches used. We discuss two distinctly different kinds of visualizations, *product visualization* (visualization of the simulation model's outputs) and *process visualization* (visualization of the model's processes per se). Scientific research through computer modeling has become standard practice at many government, industry, and university supercomputer centers. Though product visualization is extensively employed in these activities, process visualization is rarely used. Our thesis is that both kinds of visualization are valuable for furthering scientific insight and understanding. Further, we feel that their integrated use becomes essential as models become more complex.[1]

Computer simulation and model-based inquiry are valuable in science education as well as research (Richards, Barowy, & Levin, 1992). To make model-based inquiry methods accessible to students we must provide semantically transparent visual representations of both model structure and model behavior. To support the conceptual clarity required for student modeling work we have devel-

---

[1]Some models are becoming so comprehensive that no scientist is expert on all aspects (e.g., global warming supermodels integrating diverse interacting submodels each of which describes phenomena such as the air–ocean interface, precipitation from clouds, and ozone layer effects).

213

oped new computer tools for visualization of model processes. We describe two such tools that were designed for secondary science education. The use of the methods exemplified in these tools could also have significant benefits for science research.

MODELING AND VISUALIZATION
IN SCIENCE RESEARCH

Computers are beginning to transform the way science is done. Scientists are using computers to model complex processes of diverse phenomena, ranging in scale from the inner structure of the proton to star cluster formation and decay (Corcoran, 1991; Kaufmann & Smarr (1993). Computer modeling adds a new dimension to scientific inquiry, complementing the classical paradigms of experiment and theory. A computer model is both the concrete embodiment of a theory and a new kind of laboratory for exploration and experiment. Computer modeling can be an illuminating source of creative insights about the structure and behavior of complex phenomena that were previously inaccessible, and it has made possible the solution of problems previously thought unsolvable. Further, modeling provides a powerful bridge between theory and experiment, informed by each to guide the other in a new synergy that extends and enhances scientific inquiry. It has already made invaluable contributions to frontier research in astronomy, biology, chemistry, physics, and meteorology.

These developments have been greatly accelerated by the use of supercomputers coupled with powerful graphics display processors. Indeed, in many cases the research would not have been possible without the use of such resources. When the phenomena being modeled have high-dimensional nonlinear interactions, traditional numerical or graphical presentations of results are not readily informative. As supercomputers are used with larger and more complex models, visual presentations become essential for understanding the results of modeling runs. In current practice at supercomputer centers the end result of "scientific visualization" is to turn the numbers generated by modeling runs (the model output data) into pictures, typically in the form of computer movies. These reveal far more vividly than can numbers, the behavior of the phenomena being modeled and the effects of model processes and interactions.[2] Portraying numerical results as three-dimensional images moving through time with color encoding produces a visually compelling and highly informative presentation that greatly aids comprehension and interpretation of model output data.

There are extensive applications of computer visualization methods of this

---

[2]"The most exciting potential of the widespread availability of visualization tools is not the entrancing movies produced, but the insight gained and the mistakes understood by spotting visual anomalies" (McCormick, DeFanti, & Brown, 1987, p. 6).

kind in current science research (Cromie, 1988; FCCSET, 1993; Friedhoff & Benzon, 1991; Haber, 1990; Kaufmann & Smarr, 1993; McCormick, DeFanti, & Brown, 1987). A recent issue of the International Journal of Supercomputer Applications (Follin, 1990) describes modeling applications in climatology, planetary studies, fluid dynamics, automotive engineering, and chemistry, including the following research topics:

Effects of increased greenhouse gases on global climate

Evolution of severe thunderstorms

Oceanography of the Pacific Ocean

Spectral classification of Jupiter's clouds

Visualization of flow in computational fluid dynamics

Three-dimensional viscous flow in gas turbines

Flow through biofluid devices (artificial heart)

Automobile side member collapse

Visual simulation of a chemical reaction

Glass structures and transition

Quantum chemical molecular models

The articles include snapshots of the visual outputs of the models. An accompanying videotape shows the animations of the output data generated by each of the models. Some, like the animation of a developing thunderstorm, are vividly realistic and lifelike; some show real objects that would otherwise be unseeable.

The complex models that are run on supercomputers are usually developed by researchers on a workstation prior to high-speed "production runs" on the supercomputer facility. The key roles of the supercomputer facilities in current practice are supporting the execution of computation- and data-intensive models and producing appropriate numerical data for subsequent analysis. The postprocessing of the results of model computations to turn the numerical outputs into animated graphical presentations—pictures or movies—is what is meant by "scientific visualization" (Foley, 1990). This phase often involves the use of powerful graphics workstations.

The representation of a scientific model as a computer program is a complex process involving representations at several levels of abstraction: the conceptual entities that scientists envisage in their mental representation (i.e., the objects and processes being modeled), a mathematical description of the behavior and interactions of these objects, a computer program for implementing the mathematics, and the visual representation of the results obtained from running the model. Typically, in current scientific practice, the conceptual entities are described by differential equations, a programing language—often Fortran—is used for transforming the equations into programs, and graphical rendering and presentation software is used for transforming the outputs of the programs (the product) into scientific visualizations. It is interesting to note that the pictures generated in the last stage of processing, the product visualization, correspond more directly to the conceptual entities in the scientist's mental model than do the intermediate representations. Indeed, new languages that more directly express mental models are likely to supplant procedural languages like Fortran in the future.

MODELING AND VISUALIZATION
IN SCIENCE EDUCATION

Under a recently completed National Science Foundation (NSF) project supported by the Applications of Advanced Technologies Program of the Education and Human Resources Directorate,[3] we developed new computer modeling tools and paradigms and explored their educational applications. We are continuing to do research and development in educational computer modeling. Our interest is motivated by several considerations. The arrival of affordable personal computers with the computational power of present-day supercomputers is imminent. Hardware systems with the capabilities of today's supercomputers will become widely available to schools during the next decade. We must start to think now about how they should be used. We need to develop the appropriate ideas, software tools, learning activities, and exemplary demonstrations.

The quantitative improvements in performance embodied in these new machines make possible *qualitative* changes in the nature of computing. These changes can be exploited to provide enormous educational benefits. In particular, real-time interactive models with richly animated graphics displays, the same kinds of tools being used with great benefit in science research, can be made accessible for use by students. The models and the modeling tools students work with will be a great deal simpler than those used by scientists, but the fundamental character of the modeling activity will be the same, as it should be. The current school science course focuses on teaching *about* science. Instead, students should be *doing* science. The way is open to introduce modeling into schools as a compelling new paradigm.

Computer modeling is valuable for students for very much the same reasons that it is for researchers. It enables students to observe and study complex processes as they are run, and to "see" phenomena that are not accessible to direct observation, thereby enhancing their comprehension of underlying mechanisms. It can provide insight into the inner workings of a process or phenomenon—not just about what happens, but why it happens. It enables one

---

[3]National Science Foundation grant MDR-8954751, "Visual Modeling: A New Experimental Science."

to make and test predictions, and to ask and answer questions such as "How will the model's behavior be affected by changing parameters?", "What are the critical dependencies?", and "How can one modify or extend the model structure so as to produce a specified behavior?" It enables investigation in situations where experimentation may be impractical or infeasible, and it enables the modeler to gain more information about a process than can be obtained otherwise, for example, by slowing down or speeding up time or by presenting simultaneous multiwindow views of different representations.

Computer modeling can dramatically enliven science education. It has unique capabilities for providing students compelling experiences, engaging them in active investigation, and enhancing their scientific understanding. A curriculum centered on modeling activities can foster the development of the notions and art of scientific exploration and inquiry. Modeling microworlds can incorporate powerful graphic interfaces to enable easy interaction without the need for a deep understanding of computers. It can support facilities that demonstrate concepts and that aid students in solving problems. A computer modeling approach to teaching science has the potential for motivating the interest of significantly greater numbers of students, not just the small fraction who are already turned on to science and mathematics.

Computer modeling is not new. Modeling languages and applications have been in use in education for some time. What is new is the possibility of making complexity more comprehensible and accessible to students through the use of state-of-the-art science modeling software with multiple visual representations of model outputs.[4] Visual simulation and modeling can greatly aid students in understanding the complex dynamic behavior of systems composed of interacting subsystems, for example in studying reaction–diffusion equations in stochiometric chemical interactions or the dynamics of competition, predation, and adaptation in multispecies population ecology models. However, modeling tools that do not support process visualization in a comprehensible representation have proved ineffective in helping students gain insight into the mechanisms underlying the behavior of complex systems. Student difficulties in acquiring model-based reasoning skills are discussed in Roberts and Barclay (1988), Tinker (1990), and Richards, Barowy, and Levin (1992).

These difficulties would surely be exacerbated by the additional complexities of models of phenomena involving concurrent interactions among multiple processes. Yet, parallel processes are ubiquitous. They are fundamental in nature, in the phenomena of physics, chemistry, biology, and psychology at all levels. They are essential components of complex interactions in everything that's interesting

---

[4]One such system is the Explorer Software science series developed by Logal Educational Software, Israel, and BBN Laboratories, Cambridge, Massachusetts, and marketed in the United States by *Wings for Learning*. Classroom work with Explorer models in mechanics and electricity is described in Richards, Barowy, and Levin (1992).

to us. Real objects in the world function, dependently and interdependently, "at the same time." In chemical reactions different processes occur simultaneously, and success depends critically on timing. Biological models of growth and change require that different processes occur continuously and synergetically. Adaptive systems—the brains of animals, ecological systems, and social organizations—are intrinsically parallel. We need to make the underlying principles more transparent and comprehensible. We need visual modeling paradigms with better visual representations for thinking about parallel processes and complex systems.

Notwithstanding its great utility, our experience is that visualization of model *outputs* (product visualization) does not go far enough in promoting students' understanding of "why things turned out the way they did." Even when the output of a model run gives a complete picture of the behavior of the modeled system, an understanding of how the system operated to give rise to the results may not be evident. Moreover, the behavior depicted may be partial and incomplete in important aspects, even when it looks right. This is analogous to a classic situation in microbiology specimen analysis. What is salient in the microscope display of a sample visually enhanced by staining may depend critically on the particular staining reagent used. Another reagent may reveal significantly different and informative features. The microbiologist who fails to take that into account may make incomplete and even faulty inferences about structure and underlying mechanism.

There is another, more insidious problem with product visualization. Visualization techniques such as three-dimensional rendering using shading and color, together with smooth animation, can produce an illusory world that is visually compelling and can seem so real that it threatens to overwhelm and hide the simplifications and defects inherent in any computer-based model of the real world. The results of even quite simplistic models can take on a superficial credibility, reflective more of the sophistication and attractiveness of the display technique than of the model itself. This is especially a concern for education. Students may well be led astray by faulty visualizations, particularly when these are coherent and seem beautiful. (This is not likely to happen with scientists, at least to those who are experts in the domain being modeled, though it is a potential problem in the context of work with supermodels, in applications integrating several complex models where there are strong interactions among the constituent models and where no scientist is an expert in all the areas modeled.)

We believe that the model development process and the analysis and interpretation of model run data can both be greatly facilitated by the introduction of process visualization tools to complement those for product visualization. As our example illustrate, these tools support animated displays of the model processes themselves, that is, the submodel structures and algorithms and their interactions during a run. They are intended to provide a visual isomorph of the student's (or scientist's) conceptualization of the model, so as to show the objects being modeled and their interactions in as direct and transparent a way as possible. The following sections describe two software systems that support process visualization: a high-level visual programing environment and a domain-specific modeling microworld.

FUNCTION MACHINES: A VISUAL PROGRAMING LANGUAGE

One of the process visualization tools we have developed is a visual programing language called Function Machines (Feurzeig et al., 1994; Wight, Feurzeig, & Richards, 1988). Function Machines uses two-dimensional iconic representations of programs, in contrast with the familiar (one-dimensional) textual languages in almost universal use today. Our primary objective in developing Function Machines was to make programing easier to use for mathematical exploration and inquiry. In working with educational programing languages, even with accessible languages like Logo, students and teachers often have difficulty understanding control structures and acquiring fluency in the use of iteration and recursion, which are central for the description of algorithms. These conceptual barriers to acquiring a nonsuperficial level of programing competence have largely been eliminated in Function Machines.

In Function Machines the central metaphor is that a function (or procedure or algorithm) is a "machine" (displayed as a rectangular icon with inputs and outputs). A machines' data and control outputs can be passed as inputs to other machines through explicitly drawn connecting paths. Any collection of connected machines can be encapsulated under a single icon as a higher order "composite" machine; proceeding in this way, machines (programs) of arbitrary complexity level can be constructed. As machines are activated and run, their icons are shown in inverse video, and the passage of data into and out of machines is shown by animating the data and control paths. Thus, the operation of a Function Machines program is visually explicit and very easy to follow. A brief look at Function Machines follows, to show the visual representation of algorithmic processes within this paradigm, and its use in mathematical modeling. The Function Machines language and Function Machines programing are described more fully in Wight et al. (1988).

The lower left side of Fig. 8.1 shows a function machine that computes the logistic function, $f(t) = \mu t(1 - t)$. The machine contains two input "hoppers" shown at the top of the window (one for the parameter, μ, and the other for the argument, t, as the labels under the corresponding hoppers indicate.) The hopper for μ is empty because it has not yet received a value; the hopper for t has its current value of .5. The Logistic machine has a single output "spout" (shown at the bottom of the window) for receiving the result of the calculation. The inside

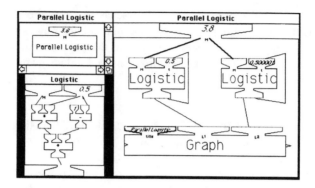

FIG. 8.1. Process visualization: The parallel logistic model.

of the Logistic machine contains three simpler machines: two multiply machines, denoted by "*", and a subtraction machine, denoted by "−".[5] When numerical values for μ and t are supplied to the Logistic machine's hoppers, it passes them to its internal machines, which perform their indicated functions to carry out the computation. Data are moved from hoppers to machines and from one machine to another (i.e., from spout to hopper) along the connecting lines shown (called "pipes"). The result of the logistic calculation is sent to the Logistic machine's spout.

The process of building more complex machines out of simpler ones can be continued to higher levels of embedding and encapsulation. The upper left window of the figure shows a composite machine, the Parallel Logistic machine, which is composed of composite machines. It has a single input, μ (whose current value is 3.8), and it has no outputs. The inside of this machine is shown in the right window. It is seen to be composed of three composites: two Logistic machines and a Graph machine. The two Logistic machines have slightly different values of t in their input hoppers (.5 and .500001). Each has two output pipes; one goes to the Graph machine and the other feeds back as the next input value of t (this is a straightforward visual way of directing the simplest form of iteration, *backput iteration*, where each output becomes the next input). The inside of one of the Logistic machines is shown in the lower left window of the figure. The Graph machine produces a graph whose title is given in its first hopper, whose abscissa values are piped to its second hopper (labeled L1) by the Logistic machine on the left, and whose ordinate values are piped to its third hopper (labeled L2) by the Logistic machine on the right. In this case the Graph machine generates a scatter plot of one Logistic function against the other.

---

[5]These are "primitive" machines provided by the system as building blocks for constructing more complex machines (programs). More than 100 such primitives are implemented in Function Machines, including the logical, mathematical, graphical, and input–output operations and constructs typically found in programing languages.

When the Parallel Logistic program runs, the user can elect to show process visualizations at any of three levels of program depth: the top level only (the Parallel Logistic machine shown in the upper left window), the top level machine together with its inner body (the three-machine structure shown in the window on the right), or the Parallel Logistic machine together with its inner body and also with the low-level Logistic and/or Graph machines. In all these cases, as the program runs, inverse video is used to highlight the machines that are currently active and to show the propogation of data along the active pipes.

At the same time, the product visualization, which in this example is the scatter plot graph, is displayed as it is generated. Figures 8.2 and 8.3 show snapshots of the product visualizations, the graphical outputs obtained from running the Parallel Logistic program. Initially, the output values of the two Logistic machines (whose initial inputs differed by only .000001) are very close together, so the points generated by their scatter plot lie on a diagonal as shown in Fig. 8.2. Subsequently, however, the two Logistic outputs become widely divergent, as shown in Fig. 8.3. This will be true even when the initial values of the two Logistic functions differ by an arbitrarily small amount. This illustrates a characteristic behavior of the phenomenon known as mathematical chaos—the exquisite sensitivity under iteration of nonlinear functions (even simple quadratic functions like the Logistic) to small differences in initial conditions.

The Parallel Logistic program shows the use of Function Machines for process visualization of mathematical models defined by equations. Models can also be implemented as *object-oriented simulations* (Schmucker, 1986). Function Machines can be used for process visualization of a particular class of such models called *turtles*. These are graphic objects displayed as turtlelike icons with a location and a heading. They respond to commands such as Forward and Back (which cause them to move forward or backward along the direction of their heading a specified number of distance units) and Right and Left (which cause them to turn around their center to the right or left a specified number of angular

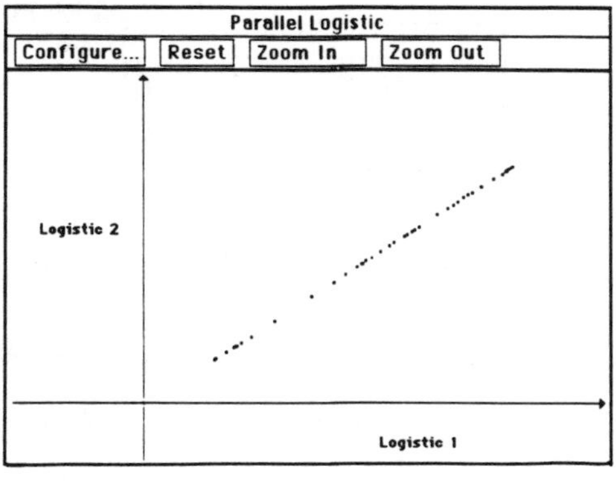

FIG. 8.2.   Parallel logistic scatter plot: Initial values.

Parallel Logistic

| Configure... | Reset | Zoom In | Zoom Out |

Logistic 2

Logistic 1

FIG. 8.3. Parallel logistic scatter plot: Chaos.

degrees). Like objects, in general, turtles have state variables (including their location and heading). They also have algorithms (called "methods" in object-oriented programing jargon) for such actions as moving forward a specified distance, turning a specified angle, and drawing their icon to show their current screen location and heading.

Figure 8.4 shows the Function Machines program and initial display for modeling an interactive multiturtle simulation called Turtle Tag. The classic turtle tag problem is to describe the pattern generated by the tracks of four turtles, initially positioned at the vertices of a square, that simultaneously move in a counterclockwise direction toward their nearest neighbors. The turtle display (in the right window) shows the initial positions of the turtles. As the program runs, each turtle first computes the heading of its nearest neighbor. Thus, turtle a seeks turtle d, d seeks c, c seeks b, and b seeks a. Then, each turtle moves a short distance along its new heading, and the process continues with further rounds of seeks and moves.

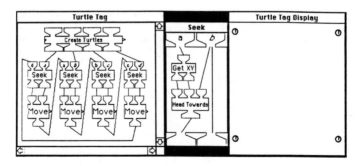

FIG. 8.4.   The Function Machines turtle tag model.

The left window of Fig. 8.4 shows the top-level program. First, the four turtles are created and given their initial locations and headings (by the Create Turtles machine). Next, the four Seek machines compute the new headings for turtles *a, b, c,* and *d,* respectively. Then, the four Move machines move the turtles forward a fixed distance along the new headings. The output of each Move machine passes the current position and heading of its turtle to the appropriate Seek machine to ready it for its next computation.

The center window of Fig. 8.4 shows the inside of one of the Seek machines, that for turtle *b* seeking turtle *a.* The Seek machine contains two primitive turtle machines, Get XY, which computes and outputs the *x* and *y* location of turtle *a,* and the Head Towards machine, which has three inputs: turtle *a*'s *x* and *y* coordinates, and the name of the turtle that is to move to that location (in this case, turtle *b*). The other three Seek machines invoke the same actions for turtles *c, d,* and *a,* respectively.

The Move machines, whose inner components are not shown, also have a straightforward structure. Each contains two primitive machines, Set Heading, which sets the heading (just computed by Seek) of the turtle that is being directed to move, and Forward, which moves the designated turtle a specified distance along the computed heading.

The right window of Fig. 8.4 shows the initial state of the four turtles, each located at the vertex of a square. As the program runs, this window will show the paths of the four turtles as these are generated, that is, it will display the product visualization.

Figure 8.5 shows the Turtle Tag program in operation. As the left window shows, the four Seek machines are ready to run. (This is indicted by displaying them in black, i.e., in inverse video.) Note that all four have been activated at the same time so they will run concurrently, in parallel. The program has been in operation for some time. The right window shows the tracks that have been generated by the turtles thus far.

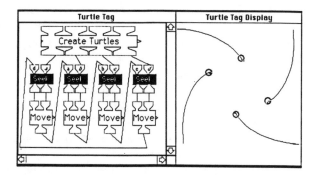

FIG. 8.5. Turtle Tag: Process and product visualizations.

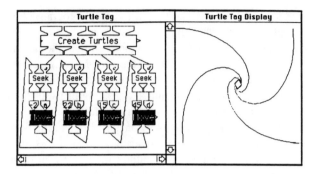

FIG. 8.6.   Turtle tag visualizations: Final state.

Figure 8.6 shows the program at a later time. The left window shows the four Move machines being invoked concurrently. The right window shows the spiral tracks that have been generated by the turtles. At this point their paths have converged—the four turtles are virtually colocated.

Figures 8.5 and 8.6 demonstrate the simultaneous presentation of process and product visualizations. As the program runs one can see the processes that are currently computing. At the same time one can also see what effects these processes have on the model's visual outputs. Moreover, one can study the relation between the program description and the program output more intensively by running the program incrementally at one's own pace, one step at a time. Observing the model processes in animation can give students very direct insight into the mechanisms underlying the model outputs. This becomes increasingly important in work with models of greater complexity.

CARDIO: OBJECT-ORIENTED SIMULATION MODELING

Although it involves concurrent processes, Turtle Tag is a relatively simple model. The turtle heading and turtle move processes, though executed in parallel, are autonomous and synchronous. Many phenomena of interest in science are a great deal more complex, often subsuming concurrent processes that have time delays, multiple feedback loops, and asynchronously coupled constituents and that, nevertheless, may have to be executed in realtime. To model such phenomena in a way that adequately captures and expresses these complex behaviors imposes computationally intensive requirements that tax the capabilities of most visual programing languages and visual modeling systems on single-processor machines.

All these characteristics of complex concurrent phenomena are realized in the Cardio modeling environment. Cardio is an object-based system expressly devel-

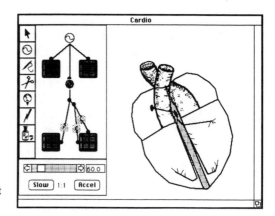

FIG. 8.7. Process and product visualizations: Cardio.

oped to model the processes that underlie the dynamics of the human heart's electrical control system.[6] Cardio provides students an interactive visual environment for investigating the physiological behavior of the heart while enabling them to gain insight into the dynamics of oscillatory processes in general (particularly coupled oscillators, which are fundamental to the operation of living systems). The Cardio program generates process and product visualizations of the heart's pattern-producing electrical control system. It enables students to investigate the deterministic heart dynamics produced by the cardiac electrical system simulation and to study the effects of changes to specific heart component parameters.

As the Cardio simulation runs, it generates several displays. The major product visualization is the *animated heart display,* shown in the right window of Fig. 8.7. This is a system-driven graphic animation of the heart model that shows in real time the rhythmic pulsation of the heart chamber. The animation is accompanied by the sound of the opening and closing of the heart valves; it is an auditory display.

The operation of the Cardio heart model is shown schematically in a second animated display, the *electrical control schematic,* which is the major process visualization of the model. The electrical control schematic is shown in the left window of Fig. 8.7. As is shown in the schematic display, there are three main types of tissue components in the heart model: pacemakers, conduction paths, and heart chamber muscles. The human heart has two pacemakers, the sinoatrial node (SA) and the atrioventricular node (AV). These are pulse generators with a

---

[6]The Cardio program was designed and implemented by Eric Neumann of BBN under NSF grant MDR-8954751. Cardio is based in part on descriptions of heart dynamics in Bratko, Mozetic, and Lavrac (1989), Glass and Mackey (1988), and Winfree (1987). The program is implemented on an Apple Mac II system.

natural frequency, but they are also sensitive to external signals that can reset them. The function of the condition paths is to conduct the electrical activations generated by the pacemakers to the heart muscles. They have associated time delays for response and recovery. The heart chamber muscles (the two atria and the two ventricles) are the target tissues of the electrical action. All three kinds of tissues are electrically excitable media. All have refractory periods for recovery after triggering. The refractory periods and time delays are the sources of the complex nonlinear behavior of the system.

The SA pacemaker node is shown at the top of the electrical control schematic display. It has conduction paths to the two atria, and to the AV pacemaker node, which, in turn, activates the two ventricles (note that it first goes to an intervening node, which has a single path to the leftmost ventricle and a dual path to the rightmost one). When the model runs, the moving electrical action potential is shown with animated trigger pulses, as seen in Fig. 8.7. The pulsing from the AV pacemaker is shown along the three conduction paths leading to the ventricles. Note that the AV node is currently active, as shown by its black (inverse video) image.

The electrical control schematic is also the user's control panel. The slider bar just below the schematic in Fig. 8.7 enables the user to increase or decrease the pulse rate of the SA pacemaker by sliding the white square forward or backward. The current pulse rate is seen to be 60 beats per minute, the normal rate. The control buttons marked "Slow" and "Accel" enable the user to slow down or speed up the model simulation rate relative to real-time operation. The ratio of simulation rate and real-time rate (displayed between the two control buttons) is seen to be 1:1 in Fig. 8.7. The icons shown at the left column of the schematic display in Fig. 8.7 represent user-selectable tools for interacting with the model. Their functions are discussed later.

Cardio also has control menus for starting and stopping the simulation; selecting a variety of output displays, such as electrocardiograms (ECGs) and phase plots; setting model parameters; and introducing abnormalities in heart function, such as blockages and rhythmic anomalies (disrhythmias.) Figure 8.8 shows the operation of the model under an extremely stressed condition. As the lower part of the left window shows, the SA pulse rate has been set to 356 beats per minute; the heart is operating at an abnormally fast rate. The upper part of the schematic shows that both pacemakers are active. The animated heart model in the right window shows a very different chamber expansion and contraction pattern from that of normal heart operation.

The heart system dynamics result from the run-time interactions of the pacemaker nodes, conduction paths, and heart chamber muscles. As the simulation runs, the interactive heart animation and electrical control displays can be simultaneously viewed with ECGs and phase plots showing heart dynamics. The left window of Fig. 8.9 shows the electrical control schematic display; the top right window shows a typical ECG plot.

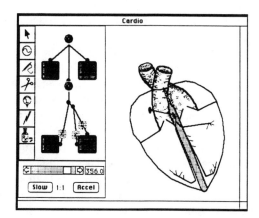

FIG. 8.8. Visualizing the effects of high pulse rate.

ECGs are useful in identifying pacemaker characteristics, conduction rate changes, and various anomalies. However, because ECGs are the result of the combined electric fields of each of the four chambers, it is not easy to visualize from ECG plots alone the complex asynchronous patterns of chamber electrical depolarization that continuously evolve over time. Abnormal patterns may arise when the heart does not return to the same state after a pacemaker completes a (normally periodic) cycle. Phase plots of the electric fields of one of the chambers plotted against those of another help visualize these complex dynamics by means of orbit paths. The bottom right window in Fig. 8.9 is a phase plot of the right atrium contraction versus the right ventricle contraction. The plot shows a

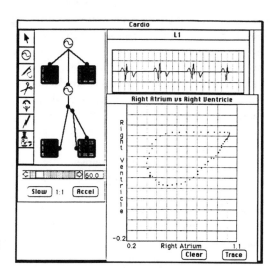

FIG. 8.9. Electrocardiogram and phase plot outputs.

limit cycle; its eccentricity depends on the phase difference between the two chambers.

To enable students to investigate the dynamics of heart anomalies as well as normal heart behavior Cardio includes files for several predefined blockages and dysrhythmias (including bradycardia and tachycardia, i.e., abnormally slow and fast pulse rates, as well as atrial and ventricular flutter and fibrillation). We have also incorporated files that simulate the physiological effects of various pharmacological agents, including caffeine and digitalis. Other types of abnormalities —such as mechanical, electrical, and chemical disturbances—can be introduced and their effects on heart behavior observed and analyzed.

The three kinds of heart components are all represented in Cardio as objects in the sense of object-oriented programing constructs (Schmucker, 1986). This helps ensure that the components function appropriately in a wide range of contexts so that, if components are deleted or added, or if the structural connections among them are modified (i.e., the heart is "re-wired"), the simulation will mirror the electrical behavior of a real heart with the same structure. Thus, students are able not only to modify the parameters of the components in a normally structured heart but also to construct abnormal heart models and investigate their diverse dynamic behaviors. For example, the effect on heart behavior from including additional pacemakers to the two normally present (the SA and AV pacemakers) can be modeled in Cardio. This anomalous behavior can occur naturally when tissue in the conducting path functions erratically by generating trigger pulses and becomes, in effect, an extra "wandering" (*ectopic*) pacemaker. Operating under the constraints posed by the intrinsic refractory limits of the conduction system, additional pacemakers yield complex echo and skip beats. Cardio's facilities enable students to compare, model, and test heart conditions like these and to investigate the complex and chaotic rhythms that they generate.

The effects of complex dysrhythmias in changing the rhythmic patterns of the atrial and ventricular expansion and contraction can be seen (and heard) in the auditory display of the animated heart. These effects, which are often pronounced, can be analyzed by studying the phase plots. For example, the student can introduce a heart block that results in a ventricle rhythm that does not always follow the SA pacemaker, thereby producing multiple orbit paths in the phase plot. This is illustrated in the lower right window of Fig. 8.10. The upper right window shows the corresponding ECG, which has a distinctly different pattern from that for the normal heart.

Students can change the parameters of any component in Cardio. They can invoke more profound changes in the model by deleting components or by adding components to form ectopic pacemakers and anomolous conduction paths. Cardio's visual modeling tools enable students to graphically create new components by selecting them from a palette, specifying their parameters, and connecting them to existing components. Thus, students can easily and quickly create their own heart models and investigate their behaviors. For example they

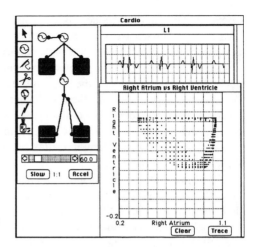

FIG. 8.10. Effects of blockage: Installing a pacemaker.

can investigate the dynamics of a heart with only a single pacemaker, to try to gain a very specific understanding of the advantages provided by the more complex double-pacemaker human heart system. (The tool used for deleting a pacemaker or a conducting path is the scissors-shaped icon in the vertical tool bar at the left of the electrical schematic diagram.)

After investigating the single-pacemaker heart students can investigate heart models with three or more pacemakers to experience the difficulties of providing control stability across a wide range of performance conditions in such models. The left window of Fig. 8.10 shows a three-pacemaker system where the third pacemaker is connected to the SA pacemaker. The additional pacemaker may have been installed by a student in an (unsuccessful) attempt to correct the induced blockage. (The tool used for installing an additional pacemaker is the pacemaker icon in the tool bar at the left of the electrical schematic diagram; the tool for installing an additional conducting path and connecting it to other objects is the pen-shaped icon just below the pacemaker icon in the tool bar.)

Students can also attempt to try to correct heart anomalies in other ways. For example, cardiac arrhythmias that have highly unstable life-threatening chaotic regimes can be simulated in Cardio. By administering appropriate electrical stimuli to the heart the arrhythmias can be controlled and converted to periodic beating (Garfinkel, Spano, Ditto, & Weiss, 1992). (The tool used for providing an electrical stimulus to any component in the model is the lightning-shaped icon in the tool bar; just above it is the meter-shaped icon, the tool for probing a component to read its electrical state.)

We describe next instances of student work illustrative of the kinds of visual modeling activities we conducted in classroom teaching experiments with Cardio. In scientific practice, computer models are often used as an essential part of research. Although the model does not usually replace the real experiment, it creates a conceptual basis for understanding it. In a similar fashion, we have designed models to behave, in a mathematical sense, like the phenomena they replicate. They can be used before experimentation to help students gain a qualitative understanding of the phenomena, and to introduce more precise and concrete language to describe the experiment. They can also be used after experiments to provide a means for examining the phenomena more carefully. But what about situations where experimentation is infeasible or impossible? The examples that follow illustrate how modeling can be used to support inquiry even in those cases.

Cardio was introduced to sophomore biology classes in the Lexington, Massachusetts, Public High School in 1990 and 1991. The discussion that follows describes the more recent intervention. There is no way, of course, that students can be allowed to perform experiments with functioning hearts in living humans. We felt it was useful, however, to make it possible for them to observe the internal operation of a living human heart, particularly those aspects relating to its electrical control system function. So we provided a substitute. Before working with Cardio, students viewed a remarkable video prepared for us by Hewlett-Packard. It showed ultrasound movies of the human heart in a cross-section view that precisely mirrors the view provided in the Cardio animation.[7]

Following that, students worked in pairs with the simulation in three 30- to 45-min sessions. In the first session an instructor described the principles underlying the Cardio model and introduced them to the program. By the end of the session they were familiar with the use of Cardio's tools and operations. They had studied normal heart function and learned how to inspect and modify pulse rate, refractory period, and the other key parameters controlling the behavior of the pacemakers, conducting paths, atria, and ventricles. They had compared the dynamics of the normal heart with those of hearts suffering from tachycardia (high pulse rate), bradycardia (low pulse rate), flutter, and fibrillation, through studying electrocardiogram and phase plot outputs as well as through direct visual and aural observations of the electrical and mechanical behavior shown in the animated heart and the schematic control displays.

During the second session, students briefly reviewed the use of the model. They were then given the open-ended problem of investigating the effects on the behavior of a normal heart as its pulse rate is steadily increased from low to high rates, and explaining (or attempting to explain) any interesting behaviors that might be observed. This is easily done in Cardio by using the computer mouse to move a slider that controls pulse rate (shown near the bottom of the schematic control window on the left side of Fig. 8.7). Students observed that as the pulse is steadily increased, the heart rate increases correspondingly (as shown in the

---

[7]The tape was prepared by Advanced Projects, Imaging Systems Department, Hewlett Packard Co., 3000 Minuteman Road, Andover, Massachusetts, 01810.

animated display by the heightened expansion and contraction of the chambers together with increasingly rapid heart sounds). As the pulse rate goes to high values, the effect becomes quite dramatic. The heart is working very hard. The loud pounding makes one feel that the heart must break when, quite unexpectedly, a pulse rate of around 300 is reached at which the heart suddenly slows down. This remarkable nonlinear dynamical behavior is due to the inability of the refratory tissue to recover in time to respond to the high frequency electrical pulse activation by the primary pacemaker. We found it interesting that several students were able to arrive at this explanation despite the fact that Cardio does not explicitly model refractory behavior at the level of detail that shows internal tissue response. They were able to hypothesize this explanation on the basis of the insights gained about heart dynamics from observations at the more aggregate level of behavior modeled by Cardio.

Toward the end of the session, students were given a more directed problem. They were shown a bradycardic heart—a heart with an abnormally slow pulse rate—and asked how they might repair it without repairing the existing heart components. The solution of adding an artificial pacemaker, shown in the left window of Fig. 8.10, was found by most students, often after proposing incorrect hypotheses about the nature of the underlying problem and exploring ineffective ways of solving it.

During the third session we involved students in peer teaching. Students again worked in pairs but one student was new to Cardio and the other was an old hand, having experienced the two sessions just described. The experienced student took charge in instructing the novice in the use of Cardio. After about 15 min of initiation, the "teacher" assigned the "student" the first of the two problems described, the "discovery" and investigation of the nonlinearity in response produced by steadily increasing pulse rate. The results of this teaching experiment were generally very positive. The experts proved to be competent instructors. The novices were well on the way toward learning to use the model to aid their reasoning about the complex dynamic behavior of the heart.

NEW VISUAL MODELING TOOLS

According to our thesis, the incorporation of process visualizations is a highly desirable addition to the technology of computational modeling, both for the introduction of the notions and skills of model-based inquiry in science education and for supporting advanced work in science research (particularly with the increasing levels of model complexity in state-of-the-art research). We have shown examples of the use of process and product visualization in two very different kinds of modeling environments, a "universal" visual programing language and an object-based, domain-specific modeling system. Both environments provide powerful tools for constructing models and for investigating model behavior. Users, both scientists and students, want to get inside the model while it is running and feel that what's on the screen is real. They want to be able to go in and make changes and see the feedback immediately.[8] Environments such as Function Machines and Cardio facilitate modeling explorations and experiments by giving users easy means to interact with the program during run time. They provide real-time access to tools for input, monitoring, inspection, display, and modification of model processes and data. We feel, however, that these tools can be further augmented to enhance the benefits of visualization and to facilitate model-based inquiry and research investigations.

As noted earlier, the use of compelling product visualizations does not guarantee that users will understand the model processes that give rise to them. Neither, however, does the addition of process visualizations (though these clearly provide a valuable new dimension of modeling power). The possibility remains, particularly with beginning students, that modeling investigations using these visualization capabilities will be no more insightful than is often the case with the simpler modeling systems. In many such "modeling" activities, students vary parameters and generate tables and graphs, but do not gain any understanding of the underlying mechanisms relating their output data to the model's inputs and actions. The existence of attractive facilities for animating the model processes and outputs does not assure, by itself, that students (or, indeed, researchers) will gain insights or understanding. We are convinced that the incorporation of constructive facilities like those in Function Machines and Cardio, which enable users to construct their own models, in addition to studying and modifying given models, are essential for fostering such insight and understanding. The possibility of assigning tasks such as "build a model with the following specified behavior" is, in our view, an essential requirement for investigating complex phenomena by computational modeling.

ACKNOWLEDGMENTS

The research reported in this chapter was supported in part by a contract from the Defense Advanced Research Projects Agency (DARPA), administered by the Office of Naval Research (ONR), to the UCLA Center for the Study of Evaluation/Center for Technology Assessment. However, the opinions expressed do not necessarily reflect the positions of DARPA or ONR, and no official endorsement by either organization should be inferred.

This research was supported, in part, by the National Science Foundation

---

[8]"Scientists not only want to analyze data that results from modeling supercomputations; they also want to interpret what is happening to the data during their computations. They want to steer calculations in close to real-time; they want to interact with their data" (McCormick et al., 1987, p. 5).

under NSF Grant MDR-8954751, "Visual Modeling: A New Experimental Science." Opinions expressed are those of the author and not necessarily those of the Foundation.

REFERENCES

Bratko, I., Mozetic, I., & Lavrac, N. (1989). *Kardio: A study in deep and qualitative knowledge for expert systems*. Cambridge, MA: MIT Press.

Corcoran, E. (1991). Calculating reality. *Scientific American, 264*(1), 100–109.

Cromie, W. J. (1988). Computer images in five dimensions. *Mosaic, 19*(2).

Federal Coordinating Council for Science Engineering and Technology (FCCSET). (1993). *Grand challenges 1993: High performance computing and communicating*. Office of Science and Technology Policy.

Feurzeig, W., Cuoco, A., Goldenberg, E. P., & Morrison, D. (1994). Special issue on function machines. In W. Feurzeig (Ed.), *Intelligent tutoring media*. Oxford, England: Learned Information Ltd.

Foley, J. D. (1990). Scientific data visualization software. *International Journal of Supercomputer Applications, 4*(2), 154–156.

Follin, S. (1990). Introduction: Special video issue. *International Journal of Supercomputer Applications, 4*(2), 3–149.

Friedhoff, R. M., & Benzon, W. (1991). *Visualization: The second computer revolution*. New York: Freeman.

Garfinkel, A., Spano, M. L., Ditto, W. L., & Weiss, J. N. (1992). Controlling cardiac chaos. *Science, 237*, 1230–1235.

Glass, L., & Mackey, M. C. (1988). *From clocks to chaos. The rhythm of life*. Princeton, NJ: Princeton University Press.

Haber, R. B. (1990). Scientific visualization: What's beyond the vision? *International Journal of Supercomputer Applications, 4*(2), 150–153.

Kaufmann III, W. J., & Smarr, L. (1993). *Supercomputing and the transformation of Science*. New York: Scientific American Library.

McCormick, B. H., DeFanti, T. A., & Brown, M. D. (1987). Visualization in scientific computing. *Computer Graphics, 21*(6).

Richards, J., Barowy, W., & Levin, D. (1992). Computer simulations in the science classroom. *Journal of Science Education and Technology, 1*, 67–79.

Roberts, N., & Barclay, T. (1988). Tools for model building and simulation in the high school. *Journal of Computers in Mathematics and Science Teaching, 8* (4).

Schmucker, K. J. (1986). *Object-oriented programming for the Macintosh*. Hasbrouck Heights, NJ: Hayden Book Company.

Tinker, R. F. (1990). *Modeling: Instructional materials and software for theory building* (NSF Final Report). Cambridge, MA: Technical Education Research Centers, Inc.

Wight, S., Feurzeig, W., & Richards, J. (1988). Pluribus: A visual programming environment for education and research. *Proceedings, IEEE Workshop on Languages for Automation*.

Winfree, A. T. (1987). *When time breaks down: The three-dimensional dynamics of electrochemical waves and cardiac arrhythmias*, Princeton, NJ: Princeton University Press.

9 Inventing Technology Assessments on Local Area Networks: An Estimate of the Importance of Motives and Collaborative Workplaces

Hugh Burns
University of Texas at Austin

In Lewis Carroll's *Alice in Wonderland* (1960), Alice experiences, if not learns, an important lesson about purpose. When she encounters the Mock Turtle and the Gryphon, she realizes how difficult it sometimes is to agree on the meaning of words. Although Alice did not assess technology in Wonderland, her adventures help illustrate the ambiguity of interpreting motives and the difficulty of reaching mutual understandings.

> The Mock Turtle said, "No wise fish would go anywhere without a porpoise."
> "Wouldn't it, really?" said Alice in a tone of great surprise.
> "Of course not," said the Mock Turtle. "Why, if a fish came to *me,* and told me he was going on a journey, I should say, 'With what porpoise?'"
> "Don't you mean 'purpose'?" said Alice.
> "I mean what I say," the Mock Turtle replied, in an offended tone. (p. 97)

Although Alice will be unable to negotiate with the offended Mock Turtle who always means what he says, the technology assessment community should work together as a community of purpose, on purpose.

Where does technology assessment begin? Why should we assess technology? What are the goals of technology assessment? By agreeing on the purposes of a technology assessment, the people in an organization will understand better what may be accomplished and what is motivating their organization to change through the power of technology. The challenge of preparing a technology assessment is, in part, to create a climate for meaningful change. To create such a climate, an organization must understand what motivates change in its people through their interactions. How can this be done?

A rule of thumb for beginning a new technology assessment should be to start with a negotiated sense of the technology assessment's product and process.

As a product, a technology assessment empowers customers to use technology appropriately and wisely. Although a technology assessment may look back, it may not dwell on the past as an evaluation must. At best, technology assessment is an artifact of a demonstrated willingness to change and a roadmap for a journey toward a better and well-motivated understanding of "what, where, when, who, why, and how" technology fits in the future.

As a process, assessing technology involves forecasting, best guessing, predicting, learning from the past experience, describing, explaining, comparing and contrasting, defining, seeking out expert opinion, being open to creative suggestions, understanding similarities while acknowledging differences, and more. In short, the process of technology assessment is inventive.

This chapter argues that the invention of a significant and successful technology assessment depends on the amount of agreement between what an organization knows and what an organization needs. The fundamental prediction is that technology assessment will demand better communication tools for insuring significant, successful, and comprehensive technology assessments.

Beginning the social construction of a technology assessment in a collaborative, networked community of designers, developers, and users, through local-area or wide-area network (LAN or WAN) technology is most promising. Certainly the technology assessment community is moving toward more negotiation. Reaching more consensus on strategic goals and methods for achieving results for clients and customers are common places for improving tomorrow's workplaces. The benefits of reaching consensus early in the process by agreeing on technology's purposes, acts, scenes, agents, and agencies establishes patterns of ownership, grassroots involvement, and potential satisfaction. It can also generate enthusiasm and a demonstrated willingness to change—not merely accept innovation. As stakeholders of the future, technology assessment partners are ready to use such emerging communication technologies and to become more effective as a result. Networked computers are promising tools to help achieve the benefits of cooperation and collaboration.

This chapter examines the following two estimates of the future and offers an illustrative, annotated case example from a group of graduate students studying technology assessment and intelligent tutoring systems at the University of Texas at Austin:

• Early in the assessment process, technology assessment methods must recover more of an organization's motives and assemble better baseline data through focused, yet comprehensive, discussions of a client's purposes, acts, agents, agencies, and scenes.

• Technology assessment practitioners and clients will use local area networks to establish their own electronic assessment community—where all are stakeholders—and thereby improve the technology assessment product through collaboration and negotiated consensus.

By better understanding the process of inventing a technology assessment in collaboration and by using local- or wide-area network tools to foster this understanding in the workplace, the full potential of a technology assessment can be realized—a comprehensive technology panorama.

THE IMPORTANCE OF ASSESSING MOTIVES: EVOLVING A TECHNOLOGY ASSESSMENT THEORY

As technology assessment matures as a discipline, its practitioners will evolve a special theory of assessment that engages and accounts for the dynamics of human motivation in the use of technology. The benefits of a theory are normally demonstrated in establishing the stability of a practice so that the practice is understandable, transferable, and worthy of being replicated. The importance of gathering motives and insuring better baseline data cannot be underestimated. Thus, a more systematic and reliable inquiry process for technology assessment practitioners could help clarify that the right issues are more consistently being addressed and stay in focus even as contexts change.

Change happens. Understanding organizational change in new technology contexts is the prime motivation for proceeding with a technology assessment. Beginning a systematic technology assessment by insuring a more comprehensive recovery of an organization's human climate as well as the necessary data is an important and grand strategy that is not to be ignored.

One of the most powerful methods for recovering motives is attributed to Burke (1969). His system for inquiry is *dramatism*. The five key terms of his theory—purpose, act, scene, agent, and agency—represent specific perspectives humans share in attributing motives (p. xv). Specifically, Burke contended that any complete statement about motives will offer answers to five questions:

- Why (PURPOSE)?
- What was done (ACT)?
- When or where it was done (SCENE)?
- Who did it (AGENT)?
- How they did it (AGENCY)?

These five questions, or viewpoints, help recover what it is known, but they also help a technology assessment practitioner predict by identifying future contexts, concerns, configurations, and requirements. By adapting Burke's ideas not only to identify but also to predict the dimensions of *what, when, where, who, how,* and *why,* the earliest data-gathering procedures will exploit the potential power of this systematic, predictive theory.

Evolving a special theory for recovering what is known and what is not known is not necessarily a new idea (Aristotle, 1954; Burns & Culp, 1980; D'Angelo, 1975; Kinneavy, 1971). Attributing Aristotle's classification of causes as the impetus for his ideas, Burke traced the evolution in systematic inquiry through both Aristotle and Aquinas:

> The most convenient place I know for directly observing the essentially dramatistic nature of both Aristotle and Aquinas is Aquinas' comments on Aristotle's four causes. . . . In the opening citation from Aristotle, you will observe that the "material" cause, "that from which (as immanent material) a thing comes into being, e.g. the bronze of the statue and the silver of the dish," would correspond fairly closely to our term, *scene.* Corresponding to *agent* we have "efficient" cause: the initial origin of change or rest; e.g., the adviser is the cause of the child, and in general the agent the cause of the deed." "Final" cause, "the end, i.e. that for the sake of which a thing is," is obviously our *purpose.* "Formal" cause ("the form or pattern, i.e. the formula of essence") is the equivalent of our term *act.* . . . We can think of a thing not simply as existing, but rather as "taking form," or as the record of an act which gave it form. (p. 228)

Today's new technology assessment practices are certainly not harmed by acknowledging a place for classical inquiry and a philosophical foundation from which a technology assessment takes its form. In estimating the future, it is well to re-remember and appreciate the golden age of Western civilization.

A special theory for estimating the future of technology assessment begins with an appreciation of the complexity of our humanity, and more specifically, the complexities of an organization as defined and perceived by the members of the organization. Future-oriented brainstorming sessions recovering purposes, acts, scenes, agents, and agencies are just the right place to begin a technology assessment. People are never completely content for long with technologies, with either their form or their function. Therefore, there is always some motivation for purposeful change and evolution. Asking questions openly in the early stages of a technology assessment may help in discovering the directions for the changes and setting the pace for the proposed technology integration.

Assessing Motives of Purpose

What are the motives of purpose? Two come to mind: better consequences and rationales.

Better Consequences.   Especially in this age of rapidly changing technologies, ambiguous human–machine interactions, and emotionally charged political and economic contexts, technology assessments should describe hopeful consequences. Questions might include (a) what activities will be empowered by new technology? and (b) what sets of activities will be changed by technology? An assessment of consequences allows users to know what the technology could be good for and what differences it could make. By foreshadowing consequences, the technology assessment should address how faithfully change could satisfy the desired human agendas.

Rationales.   A technology assessment should begin with a full, if not complete, understanding of why changes may be necessary or desired. An organization should clearly identify the "why?" and "why not?" points of view. Topics could include, for example (a) Why is this technology necessary now?, (b) What were the steps leading to the technology assessment?, and (c) Is the rationale coherent and understandable throughout the organization?

Assessing Motives of Act

The relationship between purpose and act should become evident in the process of conducting a technology assessment. The design and fit of a technology in an organization must be appropriate. Two matters of act should be assessed early in the invention of a technology assessment: technology usefulness and usability.

Usefulness.   In an organization, technology seems to have a disposition toward usefulness—features in its design that account for its function and helpfulness to the technology's users. These dispositions should be apparent—and eventually transparent—in an organization. Although it may sound like a pathetic fallacy, a technology's disposition toward usefulness allows humans to extend their imaginative reach. The following are some questions that should be addressed:

- What is happening with technology in our organization?
- What is the new form of the technology being assessed?
- How will it act in our organization?
- Will the proposed change make an organization's services more useful to our customers as well?
- As this usefulness disposition is actualized, how will this technology reveal its next incarnation?

Usability.   Technology must be usable. Generating usable technology implementation plans is an essential goal in creating a value-added technology assessment. The experience of trying to use a new technology is often intimidating. The actions in a user's head and heart are not always the precise actions the technology is taking. Important aspects to examine might include:

- How will explanations be facilitated by the technology?
- In the near term, how will technology facilitate problem-solving processes in an organization?
- How will the technology interactions promote expert performances?
- How will the technology be learned?
- How much training will be necessary?
- How much training is affordable?

Assessing Motives of Scene

Understanding the setting for a technology assessment cannot be ignored. In fact, scene issues are often the most complex and the most difficult to confront. If done well, technology assessment initiates new designs and new developmental agendas with specific settings in mind. Two reasons for assessing motives of scene are reaching a better understanding of conditions and accommodations.

Conditions.   Technology assessments must take into account an organization's climate and assess the conditions that will enable any change. Questions might include:

- Are organizational constraints imposed by facilities and by schedules?
- Do the conditions under which this forecasted change will occur account for the quantity and the quality of the technological change?
- What must change physically for the new technology to be successful?
- What must change psychologically for the new technology to be successful?

Accommodations.   The nature of value-added technology today depends heavily on the political climate in an organization. Settings often determine whether a new technology will empower all. Knowledge and technology can be constructed or deconstructed, used or abused, shared or not shared. Therefore, initial inquires should address the following questions:

- How will technology help to establish a common ground for the political, social, and cultural variables in an organization?
- How will technology accommodate the knowledge of a community?
- How well will technology be integrated considering aspects of cultures and levels of power?

Assessing Motives of Agent

The people who assess and the people who are assessed provide topics for deriving another set of motives. By examining agents, a technology assessment provides analyses for understanding—more formally—human behavior and especially the interdependence of technology and humans in performing complex tasks. Experiences and readiness are two of several topics for gaining insights into the motives of agent.

Experiences.   Experience always has been a great and grand teacher. When it comes to assessing new technologies, new ways to behave and to "improve" performance will become apparent by surveying the experiences of the stakeholders. Comprehending the potential impact of a new technology is much easier if actual experiences with a technology are available. Special experiences are especially useful because they provide a personal context for understanding whether or not a technology is desired, needed, or essential. Some starter questions would include:

- What experiences make this change possible?
- What "breakdown" experiences have we had with recent technology?
- What historical trends make our organization willing to change at this time?
- What special, new experiences are we expecting in the organization as a result of this change?
- Who has the most experience in implementing successful change in an organization?

Readiness.   One of the most obvious advantages of tomorrow's technology implementation will be simply having more prepared people. More general literacy and more scientific literacy will be necessary if we are to survive as a society. This will be true of tomorrow's smart work force. Anyone who is trying to improve performance of a team can attest that the more time spent in practice, the greater the likelihood of a successful and satisfying performance. Tomorrow's technology will be manifested in a "compression of process." This trend means that work habits will need to be revised as technology generally facilitates the capability to do more with less. The following questions are used to assess readiness, for example:

- Does technology better prepare the organization for tomorrow?
- Are the people prepared to accept technological change?
- How should an organization predict its readiness for technological innovation?

Assessing Motives of Agency

Of course, the tools themselves must be assessed. The instruments themselves contribute toward an understanding of motives. If done well, technology assessment often initiates new designs and new developmental agendas. This is a compelling attraction of technology. Foremost among the motives of agency are simplicity and propriety.

Simplicity. Keep it simple—words to the wise. An assessment of the tools will reveal how people who are not as familiar with computers can immediately start using technology. Another practical payoff might be the capability to design a technology assessment toolkit for facilitating advanced technologies. The impact of such flexible technologies is only beginning to be discussed and interpreted; studies such as Winograd and Flores' (1986) are stimulating the discussion of future design dynamics. Issues to be examined could include:

- What information tools does our organization need, desire, and appreciate?
- By what means will the technology assessment be accomplished?
- What implementation tasks will be necessary in the technology insertion process?
- How can an organization simplify its own complexity through technological innovation?

Propriety. Finally, a coherent viewpoint must also be tailored to an individual organization's or user's needs. Proposed architectures for an appropriate implementation of new technologies should be described both in terms of how that technology is understood and how it can be misunderstood by users and experts alike. The consideration of being able to judge what is the "right" technology for the specific situation makes appropriate technology assessments critical. Start with an examination of issues such as:

- What is the most appropriate use of technology in our organization?
- What is the right level of sophistication for an organization in light of the new technology?

THE IMPORTANCE OF NETWORKING: EVOLVING COLLABORATIVE ASSESSMENT PRACTICES

Vallee, Johnansen, and Spangler (1980) asserted, "A scant 100 persons throughout the world now use computerized conferencing on a regular basis. But the time may be fast approaching when far more people will be conferring through computers and we will begin to view computer conferencing as a 'natural' way to interact" (p. 298). What an excellent estimate of the future they made! The "fast approaching" prediction was absolutely true.

Today, millions of people use networked technology to transmit electronic mail and to communicate. Vallee also used terms such as "invisible colleges" (p. 298)—communities of adult learners linked together by telephone lines and transmission of information. He cited such advantages as "fast thinking that would enhance our collective abilities to resolve conflicts, deal with crises, or improve decision-making capability" (p. 299). He and his colleagues also pointed out how useful "transcripts would be as threads or chains of thought were created and later evaluated" (p. 294). Zuboff (1988) also presented and predicted many scenarios for portraying the dynamics of transforming information models to knowledge models. All these predictions are coming true. Now these emerging truths could become common places for predicting the future of how people will accomplish technology assessments in a modern workplace.

The success of a technology assessment obviously depends on how the overall communication plan is conceptualized and implemented. The more revealing the process becomes, the better the product. The newest generation of computer networking software is a dynamic medium in its own right, vastly different from books, film, and television. Tomorrow's communication software should be designed to allow users to interact exponentially. The group communication tools in a local network provide a software-driven forum for investigating, exploring, and stimulating the "knowing" processes. Designs for technology assessment tools are moving toward user-centered, collaborative environments. LAN solutions may help such users by working in close collaboration, thus being able to negotiate the meaning and the construction of the assessment parts together. Because such a major increase in information is certain to create information abundance, transacting the business of technology assessment will come to be viewed as a whole (though not necessarily indivisible) process.

The advantages and disadvantages of a dynamic environment for constructing texts, for example, technology assessments, will change the nature of human discourse itself. Fresh concepts, such as hypertext and hypermedia, will emerge as most important attributes of an evolving set of advanced technologies, as Barrett (1988, 1989) illustrated. Tomorrow's networked tools permit more opportunities for real-time communication as well as community-controlled on-line conferences and discussions, electronic mail for extending the discourses, and even off-line communications.

Beginning to view computer conferencing as a "natural" way to interact and to invest and construct technology assessments will focus attention on the widespread communication. Networked communications will enliven the assessment process. Technology assessment practitioners and clients will use local area networks to establish their own electronic assessment community—where all are stakeholders have a voice. The technology assessment product will be improved through collaboration and negotiated consensus.

Discovering the social patterns and practices of networked communication are certain to influence the precision and transfer potential of future technology assessment practices. Such understandings will allow more reliable methods for reaching consensus and, thereby, more valid instruments for predicting a technology's efficiency and effectiveness. These trends are unmistakable.

THE IMPORTANCE OF VOICES: EVOLVING REALTIME TECHNOLOGY ASSESSMENTS

The following sample of a collaborative technology assessment was conducted at the University of Texas at Austin in the College of Education's Learning Technology Center (LTC). In the LTC, various methodologies and curriculums are designed and developed for real-time learning environments. In this computer-assisted research setting, technology assessment tools and techniques are being developed for documenting an "electronic technology assessment community at work." The general research and development in electronic, collaborative learning investigates issues in creating a computer-assisted learning environment in a variety of courses and curriculums.

The particular computer-assisted learning environment allows for electronic mail, word processing, prompted question answering, peer-assisted revision, and real-time electronic class discussions. This innovative environment was designed to be richly collaborative and cooperative by The Daedalus Group. Originally, these tools and software features were designed and developed for managing intensive writing instruction on a local area network (Bump, 1990). Daedalus™ Interchange integrates collaborative, groupware functionality in a system that allows for many-on-many interactions. Daedalus™ Interchange is the real-time or synchronous tool in the integrated communications environment.

The following transcript includes excerpts from the first group interaction on the process of understanding what a technology assessment is. The group was specifically interested in issues of technology assessment policy and methodology, especially an assessment of intelligent tutoring systems in an organization. These selected excerpts illustrate some of the issues that were examined. It is important to realize that all of the participants are writing, reading, and responding at the same time. The laboratory is set up so that 14 participants can collaborate simultaneously. The students are enrolled in a graduate course in intelligent tutoring systems. All of the participants have prepared by reading Baker (1991; see also Burns, Parlett, & Redfield, 1991). Their willingness to explore these evolving tools, not to mention their ready wit and wisdom, is appreciated. The number or numbers in the brackets refer to the chronological sequence of messages. With the exception of my name, all of the participants' names have been changed. These excerpts have been edited for typographical errors and spelling.

Hugh [1; 2; 3; 4]:

Please join the Technology Assessment Conference in order to join in our InterChange on technology assessment. . . . What is technology assessment and what is its purpose? . . .

Eva L. Baker from UCLA . . . stated: "The concept of technology assessment should be substituted for evaluation in our thinking" [Burns et al., p. 245]. What does she mean? . . .

244

You may choose to start by looking at the characteristics of technology assessment—the indicators, the technical issues, the criteria, the impact, the costs, the procedures, the expert analyses vs. the empirical analyses. . . .

Or: How about those organizational issues? But here's the catch for some of you. What issues in technology assessment do you feel Baker may have left out—assuming you buy her argument that technology assessment is LARGER than evaluation (see responses to the first question).

I began the session by announcing the discussion topic for the session on the local area network. In terms of foreshadowing how the class will respond in this environment, I also mention that we will be interested in watching how Baker's thesis is supported. In summary, this discussion opens with a statement of purpose, prompting questions about a researcher's agenda. Matthew responds first by composing a short, abbreviated summary of Baker's major points. Then others contribute their initial definitions of technology assessment just to get started:

Matthew [5,6]
"Assessment" looks forward. Determines current situation and projects future possibilities. "Evaluation" looks back. Determines past performance. Four reasons for substituting assessment for evaluation:

    1. E[valuation]. Offers narrow methodological choices.
      a. empirical collection of data (control groups).
      b. quantitative analysis as basis for inference.
      c. focus on summation and reduction of data.

"Viewing evaluation as bound by constrained methodology is a widespread misperception" [Baker, 1991].

    2. E. divided into two parts:
      a. summative—used to make decisions.
      b. formative—used to improve programs.
      These are not necessarily mutually exclusive.
    3. Most E. uses limited criteria for judgment.
    4. E. usually deals with individual, specific cases.

Thus, the results have no generalizability. Baker suggests using the term "assessment" because it has a more global, broad-based connotation. This better suits the needs of determining and predicting the impact of technology than the traditional images conjured up by the term "evaluation."

Charles [7]
Technology assessment implies a proactive look at the benefits/dangers of a given technology—not merely an evaluative approach of looking at what has been done.

Priscilla [8, 9]
In thinking about Baker's definitions on assessment and evaluation, she seems to feel that evaluation is a static element while assessment is an evolving element. I

will try to put this into an example in my own words. If you went on a trip and arrived at a destination, you might stop and ponder your situation. If you thought about where you had come from, what had happened to you on the way, and how you benefitted from the trip, this would be "evaluation." However, if you considered where you were going tomorrow, made plans on what to see and do, and considered possible obstacles that might occur, this would be "assessment." . . . I think that assessment helps prepare you for the future by making predictions and by considering risks. However, I think a good and valid assessment has to involve evaluation. It brings to mind the old saying about the need to study history, otherwise the same mistakes are repeated.

Maria [10]
As Matthew suggested, according to Baker, assessment looks forward with the ability to plan and predict future risks. Evaluation, although operationally similar, looks back on what has already occurred. In order to elaborate on the differences between assessment and evaluation, Baker cites how two different government organizations use information for decision making. Citing the Office of Technology Assessment, Baker suggests that the role of this organization primarily revolves around policy making, a forward-looking activity. In contrast, the General Accounting Office uses information to justify the allocation of resources, which can be perceived as more of a review process. In the context of these two organizations, policy making because of its predictive nature requires assessment; while justification, a review process, requires evaluation.

Charles [13]
Baker feels that evaluation came from the old social science school, and has seen its day. For example, social science (in the early days) began with models from the natural/physical sciences—these models didn't work well for certain human variables (especially continuous variables like intelligence). Baker believes that quantitative models don't necessarily make for good inferential decisions.

Miranda [14]
Baker asserts that evaluations lack generalizability because individual instances are addressed, usually not very thoroughly. I buy into her argument for supplanting evaluation with assessment. Technology is NOT simply another delivery system to be "evaluated." Technology IS "interactive, dynamic and develops rapidly. . . . "
My FAVORITE part: "Technology is almost guaranteed to generate . . . outcomes and applications that were previously not considered by the training system nor imagined by the tech designer." This comes from the anthropologist in me. People adapt technology to uses in spite of it all. THEREFORE a "critical element in technology assessment is identifying when these options represent powerful, useful approaches, goals or recombination or redefinition of prior goals."

Hugh [15]
Great analogy Priscilla—the travel sequence! Seems that technology assessment means reading the tea leaves or reading a horoscope. How can we make technology assessment ever work if it takes place in the future?

Sarah [23]:
Technology assessment is about taking the time to develop a short range and long range understanding of the impact of new technologies. Baker's article is very comprehensive in defining the assessment, contrasting it to evaluation, discussing the range of technologies to be assessed, criteria for selecting technologies to be assessed, etc., but what is missing? The methodology for making this time and information intensive process manageable. Also, a scaled down version so that entities can use technology assessments to make lower level business decisions.

In the initial sequence of this technology assessment, the participants are establishing a common ground for understanding the purposes of a technology assessment. As the transcript continues, the group on-line conversation becomes more specific and more tailored to the participants' experiences. Each participant's potential applications of technology assessment is a bit different. With a tentative consensus established and an operational definition in place, more and more individual differences begin to distinguish the group interaction. The analogies, for example, help to distinguish the individual viewpoints and observations.

Priscilla [35]:
One of the most impressive points made in the article was the following: "The initial effects of the technology are almost always underestimated. Studies of technology must be especially sensitive to the notion of technology-push—for technology is almost guaranteed to generate, by its very existence, outcomes and applications that were not previously considered by the training system, nor imagined by the technology designer." This thought is mind-boggling to me. It says to me that a designer of technology never creates a finished product because it is always being expanded upon, modified, and improved. It would be like an artist making a few brush strokes on a canvas, then someone else coming behind and adding a new color or design. Following him would be a line of artists adding their own particular touches as the picture would grow and change. The picture will never be finished, but continue to expand and each viewer would see a different aspect of the picture and appreciate it in different ways.

Hugh [38]:
Priscilla, you are on a roll this evening. The maturing of a technology does make it set in its ways—especially when the investments have been expensive in terms of money and human resources, e.g., training costs. Can you imagine having to reconfigure the training technologies for the Space Shuttle. We had better be about the technology assessment of the Space Station Freedom, wouldn't you say?

Charles [39]:
Priscilla, I sure like your canvas analogy—seems to me that this is the way computing (for ex.) evolves. It really is the layers upon layers that makes for a rich technology—and none of the layers can be omitted—even need that touchy-feely soft science stuff.

Matthew [41]

Dr. Burns, You do seem to have a point there about tea leaves and horoscopes. Baker says that tech. policy makers must accept a "period of suspended disbelief" (p. 249) because assessment of the impact of a certain technology, especially the results that were not specifically designed or predicted, takes a long period of time and a wide range of tests that cover a number of indicators. Because of this, Baker feels that assessment should be carried out by an institution that is dealing in the long term and not by an individual or individuals who see the assessment as a short-term project.

Hugh [42]:

Just curious. How many of you have ever been on a technology assessment team? How many of you have ever been on an evaluation team?

Priscilla [44]:

Baker makes the point that policy makers should "accept a period of suspended disbelief" concerning new innovations and ideas for technology. Basically, that says to me that one has to be open to the possible (no matter how impossible they may seem at the time) new uses for technology. Remember how Dick Tracy has a "TV watch" back in the 1940's—that type of technology is now possible. Is it because designers keep their options open in "suspended disbelief" until the technology matured enough to create something that an artist has conceived in his mind?

Charles [45]:

I've done evaluation.

Hugh [46]:

Sarah, I thought you might be the one to discuss what is missing in Eva's article—considering that you have brought up all those ethical and sociological matters a couple of weeks ago. Dr. Baker has allowed us to see just the tip of the this iceberg called technology assessment.

Sarah [47]:

Baker says on page 248 that "technology assessment attempts to determine the full range of use." Also, Baker cites on page 245 the deficiencies of performing a technology evaluation. However, where are the benefits listed? And what is the clearly articulated purpose of the technology assessment. Based on the group's comments, we all acknowledge that predicting the full range of uses for a technology is a next to impossible undertaking. The technology assessment described here seems too unwieldy, too murky, too involved to ever be performed. And if one is performed, are many? And how far short do they fall of Baker's very theoretical approach?

Mark [50]:

Miranda, your comment about the way people actually use technology reminds me of an often quoted sentence from William Gibson's cyberpunk novel—Count Zero: "The street finds its own uses for technology." What is important about E.B.'s [Eva Baker's] idea of assessment of the technology is that it actually strives to investigate the unforeseen benefits and dangers of the technology which were not intended by the designers. "For technology is almost guaranteed to generate, by its very existence, outcomes and applications that were not previously considered by the training system, nor imagined by the technology designer. These new uses may be described mistakenly as side effects, when, in fact, they may be the delayed but central questions of the innovation." They may even become a selling point for the technology, much like the "spin-offs" from space technology have been used to gain funding for NASA. What kind of spin-offs might we anticipate from an ITS? Careful, Nintendo may be listening.

Priscilla [51]:

In response to your question about being on assessment/evaluation teams: I have been on evaluation teams as a secondary teacher evaluating instructional programs.

Miranda [53]:

This talk about summative evals makes me digress to ideas of TQM where revisions and change seem built in to the system. If feedback is being generated from the consumers or users, then change is constant. . . . My interpretation of total quality management is eternal upgrading. . . . How 'bout it ALICIA, SARAH and GRACE?

Sarah [55]:

This canvas analogy and any other similar idea are all related. Man did not all of a sudden perfect the methods of performing liver transplants, but only after work was done in anesthesia, antisepsis (?), and on and on were we at a point to perform a successful liver transplant. We've always built on what others before us have done . . . unless as Priscilla warns, we fail to learn from history.

Alicia [56]:

Personally, I was attracted to Baker's technology assessment concept because of the futuristic outlook. . . . As you work with the technology, you may want to document unexpected results of the teaching technology so that the shelf-life of the product can be extended. You may want to look at how the technology impacts the learner or how it facilitates collaborative learning. You may want to look at the product's role in your curriculum. If it was intended to be targeted to a particular skill and it turns out to be multipurpose, you may have just added value to the product. Have you heard about . . . hammer theory? Give a kid a hammer and he bangs on everything. Use the hammer theory in your assessment as well. Perhaps the teaching technology will not solve your intended purpose. In your expansive review of a teaching technology, look at your product as a hammer for awhile. See what it hits well and what it misses.

Priscilla [59]

Back to the idea of "suspended disbelief" for technological innovations. With the high cost involved in developing these systems, the nation in a recession, and having a 4-trillion dollar debt, how much leeway do we have in spending large amounts of money for all types of innovative programs. Since we can't see the future and know which technologies will spawn great advances, and which will be obsolete duds, who decides where to spend the money? Since new technology develops so quickly, it is difficult to either evaluate or assess where to go next.

Money should be put into technology, but in what areas, how much, and for what purposes?

In these selected passages, the computer-assisted group discussion created examples and simultaneously asking questions of one another. The participants' analogies have moved from an artist's canvas to hammers and back again. The group discussion becomes even more inventive in the closing moments. Such topics as technology assessment practices, on-line documentation, bad outcomes for well-intended evaluations, and resource allocation dominate the group's end-game discussions. Notice how the entries have evolved from description to more narrative. Notice also how the conversations continue well after the "teacher" drops out of the session.

Sarah [60]:
Dr. Burns, tonight: forget theory. I just want something that works. Again, what is missing from Baker's methodology is practicability. Also, by the time her assessment is performed, the technology would have evolved thereby changing some of the input that went into her assessment. I do have one objection though. She says on page 257 that expert analyses should be performed until "some" kind of consensus is reached. If we are talking tea leaves, I find this a bit flaky. What kind of consensus can be reached if they don't all have the same crystal ball? And we know that they don't and they shouldn't. One group has the ability to see the future? And what if their assessment was taken seriously. Consensus like that shouldn't be vested with too much power, should it?

Hugh [74, 75]:
Sarah—on your consensus reaching comment—good observation. There has to be some commitment to a group viewpoint about what the technology shall achieve. Then, I would guess you have to lay out a series of benchmarks so that you can know where you were, where you are headed, where you wanted to be by that moment in the future (beware the underestimates), and so forth. I am working on more practicality, I promise. Next year? . . . Well gang, time is about up for this brain-drain on technology assessment. Good session. Lots here for those of you who wish to write a response essay on this topic.

Priscilla [76]:
Matthew, how does one "detect the unimagined." How can one imagine what is unimaginable? And who makes these decisions? I may apply for that job. Would it be best to have these people with this job come from "inside" or from the "outside"—remember the problems with established technology not seeing new uses.

Sarah [77]:
Miranda, just for you, a real live response: What was the question? Ha. Umm, TQM as eternal upgrading? Well, it depends which slant you want to take. Continuous improvement could be one component of the TQM craze, but we're more familiar with the emphasis on self-managed and empowered teams and then with the statistical analysis of operations. Hey, I hear the TQM craze is winding down. Get ready for the next fad.

Maria [78]:
Alicia: I haven't heard the hammer idea before either. It sure makes a lot of sense, especially from a cost perspective.

Grace [79]:
TQM now has its own division in a government agency. . . . It will be around.

Sarah [81]:
Alicia, the 'hammer' (not M.C.) is new to me. Does it mean that take technology and think creatively? Think, can I hit "this" with a hammer and what would I get? Kind of like Dr. Burns' comment: Talk to anything just to get acquainted with the idea that in the future our world could be radically different (did I say could? I didn't mean that). For example, take the hammer to hamburgers, and what could it do? Like, take ITS (a technology) to hamburgers: menu selection, instruction for cooking, method to try to synthesize the meat, and on and on. . . . Is that how you would use a hammer in the future?

Mark [84]:
Well, Priscilla, your comment about the instructional technology product never really being completed brings me back to my earlier class today. We had two writers of the TENET [Texas Education NETwork] users' manual come to our Instructional Design class so we could learn about their experiences in designing the manual. Basically, these TENET trainers with no Instructional Design experience were writing this manual because there was no way for TEA [Texas Education Agency] to train the 16,000 teachers who were signing up for the service. They had to write the manual in 3 months, and they did a smash job off the seat of their pants. The biggest problem they had was that TENET itself was changing on a daily basis, adding features and services, improving the interface, etc. Well, they have just finished the second edition of the users' manual and in many respects it is already out of date, because TENET is still changing. What do they do? Well, they have the manual in electronic, downloadable form on TENET itself. This electronic text gets changed as required. The electronic text idea is pretty old, I notice that the on-line UNIX manual at UT gets continually updated, too. Woah! I think my textbook paradigm is changing even as we speak.

Alicia [85]:
I was part of a team selected to evaluate several competitive-procurement proposals for the acquisition of hardware for NASA. We used a comprehensive check list that included an item concerning the future life of the technology. I also participated in planning the field tests of the workstations and peripherals. In this particular case, the lowest bidder was not the winner.

Charles [86]:
Mark, Mr. Future brought to mind NPR's Mr. Science (now Dr. Science). Wonder if Mr. Future has a masters . . . in future!

Alicia [87]:
I agree with Sarah about the complexity of Baker's suggested assessment. Too many resources are required. Oftentimes educators cannot afford the teaching technology, much less the added resources to assess its impact. How much should Baker suggest to allocate to the budget for assessment? Twenty-five percent of the

total cost of the product? 35%? 45%? We would have to be able to allocate the expense in advance if you are a planner and budget preparer.

Mark [88]:
Dr. Burns, I think I have been on an evaluation team. A couple of summers back I edited some math workbooks for [a publisher]. I found lots of mistakes and made many comments about pedagogical differences I had with the authors. I was real proud of myself. But the folks [there] were really ticked off. I didn't know it, but the workbooks were supposed to have already been gone over with a fine-toothed comb. Maybe I was part of the summative evaluation and was giving the books a thumbs-down without knowing it. Anyway, this highlights some of E. B.'s concern about the instructional designers not doing their own assessment because of vested interest in their product.
    END OF CONFERENCE

This first technology assessment session lasted about 60 min. Thirteen people participated. Eighty-eight messages were entered. These excerpts represent only a part of the total text. In discovering collaborative standards and negotiating within a real-time community, the participants in this session began a technology assessment, reached some consensus, recovered unique experiences and memorable analogies, and, as you have read, enjoyed the experience.

INVENTING TECHNOLOGY ASSESSMENTS:
FINAL OBSERVATIONS

Networked software provides many design options for personal, highly collaborative interactions, filled with moments of becoming more of an expert, of understanding what was not understood, of being challenged to know and articulate important things in personally useful and publicly usable ways. Speculating from these collaborative experiments in recovering motives and discovering community, the future of technology assessment will certainly benefit from exploiting the various dimensions of network technology.

This chapter has presented a rationale for inventing technology assessments in collaborative settings. To review, some estimates are easy to predict:

1. Collaborative technology assessment methods and tools will rapidly evolve.
2. Computer networks will play greater and greater roles in day-to-day technology assessment activities.
3. Computer networks will become commonplace communication tools widely used by information consumers and knowledge producers.
4. The impact of such collaborative technology assessment will be seen soon in total quality environments.

5. Understanding how technology assessments are constructed or deconstructed, used or abused, shared or not shared will continue to be a matter worth careful study.

These predictions are fascinating. Recovering what is known and discovering the motives in the process of conducting a technology assessment will represent a major part of future technology assessment work. The method behind this chapter's madness has been to promote the power of collaborative work. Experts may continue to disagree about the extent to which information technology will be useful, but coming to terms with the human motives of innovation, with purpose, with organizational uncertainty, with technological dispositions in a complex workplace—such challenges face those who begin the process of technology assessment.

One fine way to begin inventing a technology assessment is to discover the motives—purposes, acts, scenes, agents, and agencies—for technological change in an organization. Another fine way to begin inventing a technology assessment is to collaborate and cooperate in the assessment process on computer networks. With such beginnings, perhaps future technology assessments will mean what they say.

ACKNOWLEDGMENTS

The research reported in this chapter was supported in part by a contract from the Defense Advanced Research Projects Agency (DARPA), administered by the Office of Naval Research (ONR), to the UCLA Center for the Study of Evaluation/Center for Technology Assessment. However, the opinions expressed do not necessarily reflect the positions of DARPA or ONR, and no official endorsement by either organization should be inferred.

REFERENCES

Aristotle. (1954). *The rhetoric and the poetics of Aristotle* (W. R. Roberts & I. Bywater, Trans.). New York: Modern Library.

Baker, E. L. (1991). Technology assessment: Policy and methodological issues for training. In H. Burns, J. W. Parlett, & C. L. Redfield (Eds.), *Intelligent tutoring systems: Evolutions in design* (pp. 243–263). Hillsdale, NJ: Lawrence Erlbaum Associates.

Barrett, E. (Ed.). (1988). *Text, context, and hypertext: Writing with and for the computer.* Cambridge, MA: MIT Press.

Barrett, E. (Ed.). (1989). *The society of text: Hypertext, hypermedia, and the social construction of information.* Cambridge, MA: MIT Press.

Bump, J. (1990). Radical changes in class discussion using networked computers. *Computers in the Humanities, 24,* 49–65.

Burke, K. (1969). *A grammar of motives.* Berkeley, CA: University of California Press.

Burns, H., & Culp, G. (1980). Stimulating invention in English composition through computer-assisted instruction. *Educational Technology, 20*(8), 5–10.

Burns, H., Parlett, J. W., & Redfield, C. L. (Eds.). (1991). *Intelligent tutoring systems: Evolutions in design.* Hillsdale, NJ: Lawrence Erlbaum Associates.

Carroll, L. (1960). *Alice's adventures in Wonderland & Through the looking-glass.* New York: New American Library.

D'Angelo, F. J. (1975). *A conceptual theory of rhetoric.* Cambridge, MA: Winthrop.

Kinneavy, J. L. (1971). *A theory of discourse: The aims of discourse.* Englewood Cliffs, NJ: Prentice-Hall.

Winograd, T., & Flores, F. (1986). *Understanding computers and cognition: A new foundation for design.* Norwood, NJ: Ablex.

Vallee, J., Johnansen, R., & Spangler, K. (1980). The computer conference: An altered state of communication? In S. Ferguson & S. D. Ferguson (Eds.), *Intercom: Readings in organizational communication* (pp. 290–330). Rochelle Park, NJ: Hayden Book Company.

Zuboff, S. (1988). *In the age of the smart machine: The future of work and power.* New York: Basic Books.

10 What Networked Simulation Offers to the Assessment of Collectives

J. D. Fletcher
Institute for Defense Analyses

INTRODUCTION

Most human activity is performed by individuals working within collectives such as groups, crews, teams, and organizational units. Activities performed by these collectives have high economic, social, and entertainment value. In their review of collectives in industry and business, Cannon-Bowers, Oser, and Flanagan (1992) reported a clear "consensus among those who study industrial and organizational behavior that work groups are the cornerstone of modern American industry" (p. 355). The salaries commanded by our professional athletes, as well as the general cessation of useful activity during major athletic events such as the World Series, World Cup, and the Super Bowl testify to the entertainment value of collective activity. Finally, the absolute essentiality of collective activity in military operations and the considerable social and economic implications of its success or failure have been a major topic—perhaps the major topic—of legends, epics, and histories.

On a less grand scale, the course of assessment research suggests a different point of view. Discussions and reviews of research on collectives invariably lament the relative absence and neglect of methodology for assessing collectives given the extensive literature on assessing individuals (e.g., Dyer, 1984; O'Neil, Baker, & Kazlauskas, 1992; Roby, 1957; Turnage, Houser, & Hofmann, 1990; Wagner, Hibbits, Rosenblatt, & Schulz, 1977). If we divide all research and development on assessment into two categories, one concerned with individuals and the other concerned with collectives, the overwhelming balance on almost any conceivable measure (dollars, people trained and practicing, pages of published research, etc.) would have to go to the assessment of individuals.

There may be some wisdom in this imbalance. Collective behavior ultimately depends on the behavior of individuals. If our assessments of individual proficiency do not predict collective success, we should probably question the validity of our measures of individuals before we seek some transcending measures of collective behavior. This point of view is supported by research both ancient and recent. McGrath and Altman (1966) found in their review of small group research that individual measures of job experience consistently and strongly accounted for collective success. O'Brien and Owens (1969) found that group members' abilities accounted for about 34% of the variance in their productivity on highly coordinated tasks. Jones (1974) found that individual effectiveness accounted for 36% to 81% of the variance in the success in tennis, football, baseball, and basketball teams. Tziner and Eden (1985) found that increases in both ability and motivation of the individuals comprising three-member tank crews significantly improved their performance of military tasks. It does not, then, seem unreasonable to attempt to account for collective behavior by beginning with the behavior of individuals.

Ultimately, however, we want to know if a collective was successful—did the business unit make a profit, the maintenance crew clear the road, the sports team win the game, the infantry company take the hill? Assessment that seeks to determine the effectiveness of collectives as entities in themselves is both desirable and reasonable. Further, feedback based on assessment of collective performance appears to be as critical to the learning and performance of collectives as it is to the learning and performance of individuals (Wagner et al., 1977). Fortunately, collective performance can be measured directly and is within our technical grasp.

An important, emerging tool in the direct assessment of both collective behavior and the individual behavior that produces it is networked simulation. Recently developed capabilities for linking simulators and simulations in real-time, interactive networks have provided facilities for collecting and analyzing data on the behavior of collectives at a level of detail and realism that is unobtainable by any other means. How networked simulation might be used for the direct assessment of collectives is the topic of this chapter. Before turning to this topic, however, a few comments on the assessment of collective behavior may be in order.

ASSESSMENT OF COLLECTIVES

The collectives considered in this chapter are groups, crews, teams, and units. *Groups* appears to be the most general term for collectives. Along with O'Neil et al. (1992), we can assume that groups consist of aggregations of individuals who have been collected together for some identifiable purpose and some known period of time to pursue individual or collective goals. Along with Glaser, Klaus, and Egerman (1962), we can assume that crews, teams, and units are subsets of groups that have (a) a definite, identifiable structure, organization, and communication pattern; (b) predefined task assignments; and (c) a product that depends on the coordinated participation of all or several individuals in the group. *Crew* usually refers to a small group that requires a great deal of communication and coordination on the part of all its members to complete its tasks. *Team* usually refers to a group that requires varying amounts of communication and coordination among its members. *Unit* usually refers to a component of a formally designated hierarchy of groups that determines and channels the flow of communication and coordination among them.

Some commentators distinguish among types of groups based on the homogeneity of their membership. This dimension did not appear to be useful for the present discussion because membership in the groups considered here is everywhere heterogeneous.

One analogy to use in beginning to consider the measurement of collectives keys on the measurement of individuals. The most fundamental and pervasive view of individual assessment begins with the basic pairing of a stimulus with a response. The stimulus may be a physical signal, such as a bell, or something more abstract, such as a multiple choice test item. The response may also be physical (keypress) or abstract (problem solution). Presumably the stimulus can be systematically calibrated and presented, and the response can be carefully measured.

Of course, the issue, even in simple stimulus–response pairings, is more complicated than it appears at first glance. What are the units of the stimulus and the dimensions along which they should be varied? What is the stimulus that the subject actually perceives and what actually brings about a response? What are the units of the response and the dimensions along which they should be observed? Is the observed response the response actually made to the stimulus? "We must decide . . . whether any physical event to which the organism is capable of reacting is to be called a stimulus on a given occasion, or only one to which the organism reacts . . . and . . . whether any part of behavior is to be called a response, or only one connected with stimuli in lawful ways" (Chomsky, 1959, p. 30). Such questions are the primordial stuff of experimental psychology and the bane of assessment.

Oddly, the assessment of collectives, so frequently neglected and so much more complicated in other ways, allows us to open these issues to observation— a little. A perennial topic in the assessment of individuals has been the role of internal, unobservable variable in accounting for behavior. The strict behaviorist point of view has been that variables that cannot be observed or measured have no place in serious attempts to account for behavior. Unfortunately, there are behaviors that cannot be explained, and, despite their undeniable occurrence, can be proven to be impossible on the basis of behaviorist principles alone (Neisser, 1967).

Current approaches therefore have shifted away from the strict logical positivism of behaviorism to the constructivist views of cognitive psychology. Human cognition is now understood to be overwhelmingly constructive. Recall is not viewed just as the retrieval of items from memory, but their reconstruction from more primitive cues acting as stimuli. Perceivers and learners are not viewed as blank slates, but as active participants using the fragmentary cues permitted them by their sensory receptors to construct, verify, and modify their own cognitive simulations of the outside world. However, the internal processes used by individuals to act on these cues and devise responses to them remain as difficult to observe as ever, and behaviorists' objections to their incorporation in theories of behavior must still be answered.

It is an interesting and significant feature of collectives that the internal processes that have occasioned so much debate and controversy in measurement of individuals are less controversial, more accessible, and to some extent more easily measured in the case of collectives. In assessing collectives, we can not only ask if stimuli elicited the "correct" response, but also if the process used to arrive at the response was equally correct.

This capability is best seen in Roby's (e.g., Roby & Lanzetta, 1958) still-current model of collective behavior, which can be viewed as a stimulus–process–response–outcome model. In this model, stimuli and responses are as problematic, but at least as well understood in collective behavior as they are in individual behavior. What is different is that processes in collective behavior are themselves performed by individuals—they are both observable and measurable. They are as accessible for assessment as they are to members of the collective whose behavior is shaped and cued by them.

Outcomes, which are an explicit component in Roby's model, could be considered in both individual and collective behavior. In Roby's model, they are separated from responses because outcomes of collective behavior are, as suggested earlier, a particular matter of interest and because they may be as subject to the vicissitudes of chance as they are to correct responding—the race does not always go to the swift, or the battle to the strong.

A severe problem with stimulus–response paradigms used to assess collectives is that the static nature of stimulus–response pairings do not match well the dynamic environments in which much collective behavior takes place. Many collectives must respond like job shops to a continually changing range of stimuli, some of which are externally and some of which are internally generated. In their comprehensive review of team training and evaluation strategies, Wagner et al. (1977) were careful to distinguish between established and emergent task situations. Tasks and activities needed to perform established tasks can be almost completely specified at the beginning. Tasks and activities needed to perform emergent tasks cannot—the situation emerges over time and in response to collective actions chosen. The activities most frequently associated with crews, teams, and units were found to be emergent rather than established, and the assessment of collectives must cover both emergent and established task environments.

Another issue in using stimulus–response paradigms to assess collectives arises directly from current cognitive approaches to understanding and explaining behavior. Just as individuals may be assessed on the basis of their skills and knowledge, so also may collectives be assessed on the same basis. The freedom with which cognitive approaches employ internal mediators in the explanations and assessments of individual behavior can be extended to collective behavior as well through the search for shared mental models to account for collective behavior (e.g., Rouse, Cannon-Bowers, & Salas, 1992; Thordsen, Klein, & Wolf, 1991).

By shared mental models, these researchers have in mind "mechanisms whereby humans are able to generate *descriptions* of system purpose and form, *explanations* of system functioning and observed system states, and *predictions* (or expectations) of future system states" (Rouse & Morris, 1986, p. 349). Direct consideration of how shared mental models contribute to collective behavior may be considered to be a recent component in assessing collective behavior. However, the inverse relationship between the frequency of certain types of communication and the quality of collective performance was noted in several reviews (Briggs & Johnston, 1967; Olmstead, 1992) of collective behavior. The minimization of these communications could only be achieved if the members of collectives share a common understanding of the situation, of what should be done, and of what will happen as a result of their actions. The mental models approach may be examining the same issue but entering through a different portal.

Finally, it should be noted that issues of reliability and validity are as relevant to the assessment of collectives as they are to the assessment of individuals. We should seek to know if any assessment of collectives is reliable in that it measures something, and we should seek to know if it is valid in that it measures something relevant to the topic of the assessment.

NETWORKED SIMULATION

Concern with collective performance is pervasive and by no means limited to military operations. However, in the military, the value of collective proficiency is at a premium, the stakes for collective proficiency are high, and interest in assessing collective behavior is intense. Much of current interest in collective behavior in the military has centered on the networked simulation technology originally developed for the SIMNET (simulator networking) project.

SIMNET was a major development and demonstration project begun by the Defense Advanced Research Projects Agency (DARPA) in February 1983, developed jointly by DARPA and the U.S. Army, and transferred to the Army in 1991 (Alluisi, 1991; Miller, 1992). Although the SIMNET project has successfully concluded, the networked simulation technology it developed and demonstrated is being applied to an increasing number and range of activities across the Department of Defense (DOD). It has been a central topic of senior review panels (such as those of the Defense Science Board and the Army Science Board) that have recommended wider adoption and use of networked simulation (Fletcher, 1992). It motivated creation of the Defense Modeling and Simulation Office in 1991, and it is one of seven core topics in the DOD technology base research and development program.

Networked simulation was originally developed for training applications and was intended to improve the warfighting performance of crews, teams, and units. The individual members of crews, teams, and units who use networked simulation are assumed to be already proficient in their individual skill specialties— they are expected to know how to drive tanks, read maps, fly airplanes, fire weapons, and so on, at some acceptable threshold of proficiency before they begin networked simulation exercises. Moreover, the commanders of these crews, teams, and units are expected to possess some basic academic knowledge and practical skills in the command and control of their collectives; that is, they are expected to know at some rudimentary level how to maneuver, use terrain in a tactically appropriate manner, provide mobility and countermobility, and so forth. The essential ingredient that networked simulation brings to the preparation of warfighting collectives is not so much initial instruction (although some can be provided) as an opportunity for accessible, frequent, and realistic practice with substantive, understandable, and relevant feedback. Thorpe (1987), its principal architect, stated: "This [concern with practice] emphasizes what we already know about how a team achieves mastery of its art, be it a sports team, an orchestra, an operating room team, or a combat team: Massive amounts of practice are demanded. There is no substitute" (p. 493).

Additionally, networked simulation provides both training and assessment applications with all the benefits of simulation. These have long been noted to include safety, economy, controlled visibility, and reproducibility (Raser, 1969). Lives can be hazarded, material destroyed, viewpoints and perspective changed, and events played and replayed in ways that range from impracticable to unthinkable in the "real world" using real equipment.

Networked simulation consists of modular objects intended to simulate battlefield entities. Typical entities are combat vehicles such as tanks, helicopters, and aircraft, but the entities may be anything relevant to warfighting, including bridges, buildings, dismounted infantry, and engineered obstacles. These entities, these simulators, may be geographically located anywhere because they are modular and autonomous and because they all share a common model of the battlefield and its terrain. In a networked simulation exercise, a tank crew in a simulated tank in Germany can call for support from simulated aircraft in Nevada because they are being attacked by a helicopter simulator located in Alabama.

Crews of simulators that are not connected to the network can still operate on the digital terrain and view all the consequences of their operational decisions—tanks can be maneuvered around obstacles, sunk in unfordable rivers, and driven off cliffs—with no requirement that they be connected to the network. More typically, each entity, along with hundreds of others, is connected to the network. If the simulated combat vehicles encounter allied vehicles on the digital terrain, they can join together to form a combat team and undertake a mission with all the problems of command, control, communications, coordination, timing, and so on, that such activity requires for its successful completion. If they encounter enemy vehicles, they can fight—engage in force-on-force combat engagements in which the outcome is strictly determined by the performance of the individuals, crews, teams, and units involved. No umpires, battlemasters, or other outside influences are expected or permitted to affect the outcome of a networked simulation engagement once it begins.

Networked simulation arises from a culmination of many new technological capabilities. Particularly important among these are:

- Computer image generation: Networked simulation depends on the generation of graphics by computer on demand, in real time, and at affordable cost. Technological advances that occurred just before initiation of the SIMNET program were essential in meetings its design goals (Alluisi, 1991). Its implication for the assessment of military crews, teams, and units is that they are free to travel and use the terrain of the electronic battlefield in whatever way is necessary to accomplish their missions. Networked simulation is free of nearly all the constraints imposed by prestored images that must be retrieved on demand.

- Selective fidelity: The notion and practice of selecting a level of realism (or fidelity as it is called in the simulation community) that is keyed in a cost-effective manner to training objectives was not an original idea in the SIMNET project. Earlier DARPA projects had attempted to demonstrate its utility (Alluisi, 1991), and it has been a basic principle of instructional simulation for many years (e.g., Thorndike, 1919). However, the SIMNET project provided a successful demonstration of this technology and convinced many current decision-makers of its importance and value.

- Packet switched networks: Use of digital data packets with standard formats and routings that are determined in real time based on network traffic and demand was developed, demonstrated, and established about 10 years before initiation of the SIMNET project in the well-known and highly successful ARPANET project (Kahn, 1977). A principal innovation in SIMNET was to apply this essential technology to networks of simulators in which visual displays generated by computer are updated on demand. Packet switching is key to using networked simulation in assessment because every action of every entity in a networked simulation engagement is captured in a network packet, all packets can be stored, or "logged" (and they routinely are), and all actions of all entities in networked simulation engagements can be recreated and analyzed by retrieving these stored packets.

- Distributed computing: There is no central operating system that controls processes and schedules events in networked simulation. All entities participate as co-equals. Each entity determines from the packets passed around the network whether or not to make any changes in its view of the world. Distributed computing was an established but relatively new technology at the initiation of the SIMNET project. The thorough extent to which it was implemented in the SIMNET project was unusual and innovative. It is possible for any entity to join the network or disconnect from it at any time, and it is possible for any entity attached to the network to capture and log all data necessary to recreate any networked simulation engagement.

A number of design principles governed the development of SIMNET technology. These are discussed by Thorpe (1987), Alluisi (1991), and Miller (1992). The following are particularly important among these principles:

- The 60% solution: High quality, high fidelity, high cost simulators have been within the grasp of our technological capabilities for some time, but at commensurably high cost. In the late 1970s, DARPA began to emphasize projects guided by the principle and observation that DOD training simulators do little good if they are not available, and therefore they must be made more accessible. Beginning with a distributed instructional system and a tank gunnery trainer, DARPA began a series of programs to develop a technology for low cost simulations (Alluisi, 1991). The SIMNET project pursued this approach. The principle in SIMNET was "develop quickly, be satisfied with good enough, keep the development cost and recurring cost low, and plan to throw away earlier than in the past" (Thorpe, 1987, p. 501).

- Object-oriented design: As described earlier, the SIMNET world is modeled as a collection of entities that interact with each other through a series of events. Each entity monitors its state (e.g., position, appearance) and broadcasts any changes in its state over the network. A firing entity determines whether its weapon hits its target. A target entity determines what damage it has suffered and what information to broadcast on the network about any changes in its own appearance. The electronic battlefield is created as a set of interacting entities rather than as a sequence of tasks, thereby simplifying and smoothing the transition between real world and simulated world. Networked simulation is a practical and significant application of the rapidly emerging principles of object-oriented design (e.g., Korson & McGregor, 1990).

- Modularity: As already suggested, the entities in SIMNET are modular, or autonomous. Each entity is responsible for maintaining its own representation of the simulated world. It does not require connection to the network to function on the simulated, electronic battlefield. Such a connection is required only to interact with other entities on the battlefield.

- Transmission of "ground truth": Every entity in the SIMNET world transmits absolute and complete truth about the state of the object(s) it represents. It is the responsibility of receiving entities to filter and even modify this information before presenting it to human participants. The local, receiving entity is assumed to be best at determining what information would most likely be received in the current environment, what might be degraded due to jamming, local equipment status, and so on, and what might be lost or unavailable in the emerging tactical situation. However, the data packets stored off the network contain absolute and complete truth about the state of the entities and are all available for use in later analyses and assessments.

- Transmission of state changes: Each entity in the SIMNET world transmits only the information needed by other entities to represent changes in its own appearance. For network operations, this practice reduces redundancy in network traffic. For assessment purposes, this practice ensures that all necessary information on changes is recorded by storing the data packets.

- Use of dead reckoning: Entities extrapolate from the states of all entities in their representations of the SIMNET world into the future. If a tank is moving at a particular speed in a particular direction, all simulators that must (locally) generate imagery showing the tank will continue to show it moving in the given direction at the given speed until they are informed by network packets to do otherwise. This practice also reduces network traffic by reducing redundancy. Reduction of network traffic is not just a useful engineering feature. It increases the number of entities that can interact on the network, and thereby increases the size and variety of units that can participate in networked simulation and whose performance can be assessed using this technology.

The following three types of entities are currently represented by networked simulation:

- Manned: The principal entities are manned, or crewed, simulated combat vehicles—tanks, helicopters, and aircraft. That is to say, all their key functions are directly controlled by humans.

- Automatic: Some entities are strictly automatic. Once established or set into motion, no human intervention (by individuals, crews, teams, or units) can affect their operation. An example of automatic entities in early versions of SIMNET technology were supply vehicles that were requested and then after an appropriate (i.e., realistic) period of time would appear at the place requested. They did not travel the terrain and therefore could not be attacked during transit in early versions of networked simulation.

- Semiautomatic: The semiautomatic entities (mainly vehicles) of the semi-automatic forces (SAFOR) are among the most technically interesting, innovative, and challenging aspects of networked simulation. These entities are indirectly but not directly manned, or crewed. Human control is required for their operation at some level above that of crews—a platoon commander controls a platoon of unmanned vehicles, or a company commander controls a company of un-manned vehicles, and so on—and the entities must act and re-act in a veridical fashion, they must behave as they would if they were manned. Humans on the electronic battlefield should not be able to tell the difference between manned and SAFOR entities.

Two display capabilities developed for networked simulation are also worth mentioning:

- Flying carpet: In viewing or re-viewing a networked simulation engagement, it is useful to be able to visit the battlefield in an invisible observation vehicle capable of traveling anywhere, providing wide views of the engagement, and keeping up with any action. This vehicle is provided by the flying carpet display. In addition to providing free flight throughout an ongoing or re-played engagement, it can be tethered to any entity on the battlefield so that it automatically follows the actions of that entity in detail.
- Plan view display: Another display developed specifically for networked simulation is the plan view display which provides a tactical overview of the action. This display provides animated tactical maps with common military symbology overlaid onto terrain maps of the engagement area. The animation proceeds in synchrony with the engagement that is being played or re-played on the electronic battlefield.

Although networked simulation was originally developed for training, it is finding applications in many other defense activities. It is being used in the design, development, and acquisition of materiel and systems; for instance, we can include different performance capabilities proposed for military systems in force-on-force engagements to determine if the difference they make in combat effectiveness is worth their cost. Once prototypes are built, preparation for proposed tests can be rehearsed and fine-tuned using networked simulation before beginning trials in the field. Quick strike missions can be repeatedly rehearsed in networked simulation using digitized terrain of the area where the military operations are likely to take place, simulation of the allied forces likely to be found there, and simulators locally positioned with the military units involved before they must be physically marshaled together. And networked simulation can be used to assess the proficiency of crews, teams, and units.

ASSESSMENT USING NETWORKED SIMULATION

The business of the military is warfighting, and networked simulation is a general purpose tool for, among other things, simulating combat. In terms used by Wagner et al. (1977), it is focused on emergent rather than established task situations. There is no training or assessment "intelligence" in the core of this technology. How it is used depends on the intentions of the user. If it is to assess military collectives, then the measurement system (in contrast to the data collection system) must be added in. The essential issue in the assessment of military crews, teams, and units is their success in warfighting. Networked simulation can provide work sample assessment of these military collectives. What sort of measurement system must be added to make this possible?

Assessment of individuals using work samples can use any one or any combination of five measurement methods (Guion, 1979). Their use for assessing collectives is discussed in the following comments, and they are listed here roughly in order of decreasing objectivity.

Instrumentation

Instrumentation can provide objective data on both stimulus events and the responses to them in assessment. In work sample assessment, it is often used to measure characteristics of the work product. Unlike other instrumented environments, all physical entities and actions on the electronic battlefield are both recorded and created by the instrumentation. Instrumentation in networked simulation is therefore complete and absolutely accurate with regard to the physical characteristics of the electronic battlefield and the entities on it. All data from this instrumentation are routinely collected and stored. A designated portion of the communications traffic in networked simulation exercises is also routinely stored—and all of it can be, if needed for assessment. Other instrumented data can be added and recorded as needed.

Instrumentation in networked simulation engagements makes relatively inexpensive and accessible some measures that are exceedingly difficult to collect otherwise. For instance, it is possible to determine intervisibility precisely in networked simulation. That is to say it is possible to establish exactly when and for how long any two entities on the electronic battlefield were visible to each other. The time between onset of intervisibility and firing of the first shot or scoring of the first hit could be critical in determining the success of military units. Networked simulation could reveal how critical these measures based on intervisibility are and how they can and should be used to assess the effectiveness of military crews, teams, and units.

The comprehensive instrumentation in networked simulation makes it possible to assess the performance of collectives by playing back their digitized performance data using networked simulation to both retrieve and display the data in whatever format is most appropriate for analyses (e.g., analog formats for data such as voice communications and movement, digital formats for instrumented measurements). These data can be played and re-played at any speed or level of detail that is needed for assessment. This approach has already been taken by the military to review military operations in an actual battle, "73 Easting," fought in the Gulf war. This battle was captured in excruciating detail and can now be replayed using networked simulation (Orlansky & Thorpe, 1992).

One problem with instrumentation of any sort is, as O'Neil et al. (1992) suggested, that it may create substantial temptations to trade validity for ease of measurement. Physical measures that are easy to make, may receive more weight and attention, regardless of their validity. The comprehensive nature of instrumentation in networked simulation makes the availability of valid measures more likely.

Direct Observation and Recording

In assessing collective performance it is frequently desirable to use both instrumented measures and human observations. In networked simulation, human observers can assess collective performance either while an exercise is taking place or any time thereafter. If observers miss a critical event the first time around, they can play an exercise back as frequently as needed, starting and stopping wherever it is necessary to assess the event correctly—or even to identify it as critical in the first place. Additionally, observers can replay an exercise using the plan view display, the flying carpet, or some combination of both to make observations from different viewing points until the best one for assessing collective performance is found. Eventually it may be possible to assess the criticality of an event by changing it and observing the effects of this change on the outcome of the exercise, but this capability has not yet been implemented. Finally, by comparing observer judgments with the comprehensive "ground truth" recorded in data packets, we can determine the consistency and relative utility of different observer and instrumented measures of collective performance, as Hiller (1987) suggested.

The clarity, detail, and precision of data recorded by human observers will vary with their training and experience, as Roby (1957) and many others since have noted. This issue may be resolved somewhat by networked simulation because it provides observers ways to compensate for what they may lack in training and experience. In more abstract terms, networked simulation provides job aids for observers that can lessen their need for training and experience.

Records and Biographical Data

Many variables relevant to the assessment of collectives are routinely recorded in institutionalized records. These variables may concern both the individuals who make up the collectives (e.g., Armed Forces Qualification Test Scores, years of service, occupation specialty) and the collectives themselves (e.g., prior range or field experience, equipment and materiel issued, length of time personnel have spent in their current crew, team, unit position). Current networked simulation carries none of these records. However, other data in networked simulation are in digital form and if these biographical records are also stored digitally they can be matched by computer to collective assessment data.

Also, the U.S. military routinely creates and stores data on the readiness of collectives. These data are drawn from ARTEPs (the Army Training Evaluation Program), from performance in instrumented ranges such as the National Training Center, and performance in noninstrumented field exercises. Access to these data, their accuracy, and their relevance remain as issues to be settled, but they exist and could be used in assessments using networked simulation. The most obvious use for records and biographical data in assessment is to provide predictions for the performance of the crew, team, or unit so that comparisons can be made between the observed and predicted collective performance. Some work of this sort has already been completed using National Training Center data and could be applied equally well to assessments using networked simulation.

Testing

In networked simulation, direct testing of individual participants might be used to supplement already-available records and biographical data on participants, but this seems unlikely to occur in routine practice. Most testing in individual assessment involves established rather than emergent task situations, and comparisons of collective assessment using networked simulation with standard individual assessment will continue to hold poorly until more is understood about the psychometric properties of simulation. Some direct testing of collectives that uses emergent, work-sample approaches might be accomplished based on networked simulation exercises specially designed for this purpose, but, again, the psychological properties of emergent task environments based on simulation must be better understood before this approach can be widely used in networked simulation.

Ratings

Ratings may be made either by supervisors (military commanders in the case of collectives being assessed in military exercises) or by the participants themselves using questionnaires or self-reports. In either case, they are used in measurement as a last resort. They are rarely based on systematic observations. Supervisors' ratings are frequently based on the vague impressions of supervisors who may never have viewed the behavior in question. Self-ratings by members of collectives are subject to the same imprecision and biases that plague self-reports in individual assessment. Still, they may be more reliable in the assessment of collectives than in the assessment of individuals because the reports made by different members of a crew, team, or unit can be checked against each other. A benefit of both ratings and direct observer reports is that they can be richer than other measurement approaches in that they can include free-form comments that capture unexpected or otherwise overlooked aspects of the emergent situation. These comments are harder to code, but they can be heuristically more valuable than other data sources.

Turnage et al. (1990) concluded their exhaustive review of collective assessment methodologies by noting that "what is clearly needed is a way to simplify the task of individual evaluators to integrate and interpret data and, at the same time, provide a uniform data base from which multiple evaluators can draw similar conclusions" (p. VI-8). Networked simulation appears to be a major step in this direction. By instrumenting the electronic battlefield and issuing in an accessible, digital format, comprehensive and absolutely accurate data on the physical characteristics and actions of entities participating in an emerging task situation and by providing powerful new display capabilities, such as the plan view and flying carpet, that can replay as often as necessary and from any desired viewpoint the behavior of all the collectives involved, networked simulation both enhances the most promising of our measures of collective behavior and makes them practicable. It provides the foundation for a measurement system that should substantially advance our assessments of crews, teams, and units in both military and nonmilitary settings.

Difficulties remain, of course. The major one of these concerns validity. For the military, everything that is not warfighting—including instrumented ranges, field exercises, computerized wargames, and SIMNET—is simulation. The validity of assessment performed in any of these settings is unknown, and we cannot call a war to check it. Even if we could, the circumstances of any battle are so unique that a single war would not do; we would need a representative sample of wars to determine validity of our assessments. This is known as the criterion problem (Hiller, 1987), but it is a small price to pay considering the alternative. Perhaps it is better viewed as a challenge for assessment specialists.

The validity challenge is not unique to the military, but it may be less intractable in civilian applications. Cannon-Bowers et al. (1992) cited examples of 21 collectives commonly found outside the military. These include quality circles, management teams, maintenance crews, product development teams, cockpit crews, surgery teams, negotiating teams, instructor teams, and athletic teams. Fire fighting teams, well-drilling crews, police SWAT teams, ship crews, disaster emergency teams, ground–air control teams, and others could be added to their list. In contrast to military collectives, nearly all of these nonmilitary collectives must routinely exercise their skills to meet real-world demands. An understanding of the validity of assessment using networked simulation would emerge over time if these collectives were to use networked simulation for either training or assessment. This understanding could be reflected back into military preparations for war-fighting and thereby help relieve the military's criterion problem. It would also be of value in its own right as the use of networked simulation for assessment of collectives grows.

This is to emphasize that assessment using networked simulation should find many applications in the nonmilitary world. These civilian applications can capitalize on the highly instrumented and flexible data recording capabilities of networked simulation as well as on its capabilities to view, and review, the behavior of collectives in emerging situations, using its facilities for massive data storage and retrieval combined with easily accessible visual display. The many collectives found in civilian workplaces stand to benefit from the widened application and further development of networked simulation for training, product development, and product testing—as well as for assessment. Its uses in both military and civilian settings are just beginning to emerge.

SUMMARY

Networked simulation offers capabilities for assessing the performance of collectives in both military and civilian settings that include any or all of the following characteristics:

- Interaction with many entities: The SAFOR capabilities of networked simulation permit the observation and measurement of collectives as they interact with a large number of entities that themselves must react in a realistic and perhaps intelligent fashion to the responses of the collectives.
- Controlled communication and coordination: Networked simulation allows communication and coordination both within the collective and between it and other entities to be closely controlled and manipulated as needed for assessment.
- Command and control of independent collectives: The original focus of networked simulation was on battle staffs—relatively small command groups that must marshal large numbers of intelligent entities into effective, mission-oriented action. Many command groups of this sort exist outside of the military. These collectives can be assessed using networked simulation either by manning the entities they are to control or by using SAFOR entities and thereby reducing the expense and administrative complications that assessment of these collectives now requires.
- Wide dispersion of participants: It may be necessary to assess the performance of collectives whose members are ordinarily widely dispersed and are only brought together for emergencies or other occasions when their performance in the collective is required. Networked simulation provides an obvious and natural capability for assessing these collectives without the expense and administrative complications this assessment would otherwise require.
- Large amounts of detailed data: Networked simulation appears to offer a new and significant solution to the problems that arise when the assessment

of a collective requires the recording and retrieval of large amounts of data collected at frequent intervals and/or in great detail.

- Accessibility of data: Networked simulation also appears to offer a new and significant solution to the problems that arise when the assessment of collectives requires making large amounts of data recorded from emergent situations easily visible and otherwise accessible.
- Emergent task situations: Networked simulation was originally designed to train collectives that must react appropriately to emergent task and/or stimulus situations. It should be equally useful in assessing the ability of collectives to react to these situations.
- Visual task environment: The image system is at the core of networked simulation. It should be of natural utility in assessing the ability of collectives to respond to rapidly appearing and interacting visual stimuli.
- Psychological rather than physical reality: Networked simulation seems especially amenable for assessing the ability of collectives to meet decision-making challenges in which their ability to visualize for themselves the critical features of the emerging situation are at a premium—rather than the realism of the assessment procedures in mimicking the physical reality of the decision environment.

This doubtlessly incomplete list of collective assessment characteristics is intended to summarize some of the promise that networked simulation holds for the assessment of collectives. At present, its promise remains just that. Most of the work needed to design, develop, and implement within this technology a measurement capability for assessing collectives remains to be done. But the technology itself now exists and is ready to meet this challenge.

ACKNOWLEDGMENTS

The research reported in this chapter was supported in part by a contract from the Defense Advanced Research Projects Agency (DARPA), administered by the Office of Naval Research (ONR), to the UCLA Center for the Study of Evaluation/Center for Technology Assessment. However, the opinions expressed do not necessarily reflect the positions of DARPA or ONR, and no official endorsement by either organization should be inferred.

REFERENCES

Alluisi, E. A. (1991). The development of technology for collective training: SIMNET, a case history. *Human Factors, 33,* 343–362.

Briggs, G. E., & Johnston, W. A. (1967). *Team training* (Rep. No. NACTRADEVCEN 1327-4). Orlando, FL: Naval Training Device Center. (DTIC/NTIS No. AD 660 019).

Cannon-Bowers, J. A., Oser, R., & Flanagan, D. L. (1992). Work teams in industry: A selected review and proposed framework. In R. W. Swezey & E. Salas (Eds.), *Teams: Their training and performance* (pp. 355–377). Norwood, NJ: Ablex.

Chomsky, N. (1959). Review of B. F. Skinner. *Verbal Behavior. Language, 35,* 26–58.

Dyer, J. L. (1984). Team research and team training: A state-of-the-art review. In F. A. Muckler (Ed.), *Human factors review: 1984* (pp. 285–323). Santa Monica, CA: Human Factors Society.

Fletcher, J. D. (1992). *A review of study panel recommendations for defense modeling and simulation* (IDA Document No. D-1161). Alexandria, VA: Institute for Defense Analyses. (DTIC/NTIS No. ADA 253 987)

Glaser, R., Klaus, D. J., & Egerman, K. (1962). *Increasing team proficiency through training: 2. The acquisition and extinction of a team response* (Tech. Rep. No. AIR B64-5/62). Pittsburgh, PA: American Institutes for Research. (DTIC/NTIS No. AD 276 429)

Guion, R. M. (1979). *Principles of work sample testing: 1. A non-empirical taxonomy of test uses* (ARI Tech. Rep. No. TR-79-A8). Alexandria, VA: U.S. Army Research Institute. (DTIC/NTIS No. ADA 072 446)

Hiller, J. H. (1987). Deriving useful lessons from combat simulations. *Defense Management Journal, 23,* 29–33.

Jones, M. B. (1974). Regressing group on individual effectiveness. *Organizational Behavior and Human Performance, 11,* 426–451.

Kahn, R. (1977). The organization of computer resources into a packet radio network. *IEEE Transactions on Communications, 25,* 169–178.

Korson, T., & McGregor, J. D. (1990). Understanding object-oriented: A unifying paradigm. *Communications of the ACM, 33,* 40–60.

McGrath, J. E., & Altman, I. (1966). *Small group research: A synthesis and critique of the field.* New York, NY: Holt, Reinhart, & Winston.

Miller, D. C. (1992). Interoperability issues for distributed interactive simulation. In *Proceedings of Summer Computer Simulation Conference.* San Diego, CA: Society for Computer Simulation, 1015–1018.

Neisser, U. (1967). *Cognitive psychology.* New York: Appleton-Century-Crofts.

O'Brien, G. E., & Owens, A. G. (1969). Effects of organizational structure on correlations between member abilities and group productivity. *Journal of Applied Psychology, 53,* 525–530.

Olmstead, J. A. (1992). *Battle staff integration* (IDA Paper No. P-2560). Alexandria, VA: Institute for Defense Analyses. (DTIC/NTIS No. ADA 248 941)

O'Neil, H. A., Baker, E. L., & Kazlauskas, E. J. (1992). Assessment of team performance. In R. W. Swezey & E. Salas (Eds.), *Teams: Their training and performance* (pp. 153–175). Norwood, NJ: Ablex.

Orlansky, J., & Thorpe, J. A. (Eds.). (1992). *73 Easting: Lessons learned from Desert Storm via advanced distributed simulation technology* (IDA Document No. D-1110). Alexandria, VA: Institute for Defense Analyses. (DTIC/NTIS No. ADA 253 991)

Raser, J. R. (1969). *Simulation and society: An exploration of scientific gaming.* Boston: Allyn & Bacon.

Roby, T. L. (1957). On the measurement and description of groups. *Behavioral Science, 2,* 119–127.

Roby, T. L., & Lanzetta, J. T. (1958). Considerations in the analysis of group tasks. *Psychological Bulletin, 55,* 88–101.

Rouse, W. B., Cannon-Bowers, J. A., & Salas, E. (1992). The role of mental models in team performance in complex systems. *IEEE Transactions on Systems, Man, and Cybernetics, 22,* 1296–1308.

Rouse, W. B., & Morris, N. M. (1986). On looking into the black box: Prospects and limits in the search for mental models. *Psychological Bulletin, 100,* 349–363.

Thordsen, M. L., Klein, G. A., & Wolf, S. P. (19⁻1). *Using a cognitive model to evaluate teamwork* (Technical Rep. No. KATR9004-191-04Z). Fairborn, OH: Klein Associates, Inc.

Thorndike, E. L. (1919). Scientific personnel work in the army. *Science, 49,* 53–61.

Thorpe, J. A. (1987). The new technology of large scale simulator networking: Implications for mastering the art of warfighting. In *Proceedings of the Ninth InterService/Industry Training Systems Conference* (pp. 492–501). Arlington, VA: American Defense Preparedness Association.

Turnage, J. J., Houser, T. L., & Hofmann, D. A. (1990). *Assessment of performance measurement methodologies for collective military training* (ARI Research Note 90-126). Alexandria, VA: U.S. Army Research Institute. (DTIC/NTIS No. AD-A227 971)

Tziner, A., & Eden, D. (1985). Effects of crew composition on crew performance: Does the whole equal the sum of its parts? *Journal of Applied Psychology, 70,* 85–93.

Wagner, H., Hibbits, N., Rosenblatt, R. D., & Schulz, R. (1977). *Team training and evaluation strategies: State-of-the-art* (HumRRO TR-77-1). Alexandria, VA: Human Resources Research Organization. (DTIC/NTIS No. ADA 038 505)

Author Index

Subject Index

A

Animated display, *see* Computer simulation
Artifical intelligence
human benchmarking and, 21, 22
Artificial intelligence
natural language and, 85, 88-95
Artificial intelligence
Turing test, 16-19
Artificial intelligence measurement systems (AIMS), 3, 4, 10
Assessment, by collectives, 255-257, 267-270
broad interaction, 269, 270
cognitive approaches, 259
controlled communication, 269
individual vs. group approaches, 257, 258
in the military, 259-269
internal processes, 258
networked simulation, 259
reliability/validity, 259
shared mental models, 259
stimulus/process—response/ outcome model, 258
validity challenge, 268
Assessment,
Individual vs. group behavior, 255-259
memory and, 258
stimulus/response, 256-258
Automation, goals of,
Army's Automated Systems

Approach to Training (ASAT) 184

B

Benchmarking,
as standard in measurement, 15, 16
cognitive skill instruments, 22, 23
computer-science driven, 15
expert system evaluation, 16, 20
GATES system, 30-42
natural language application and, 21, 22
psychological process driven, 15
see Cognition, Human benchmarking

C

Cognitive activity,
metacognition, 24-26
monitoring, 23
planning, 23
see GATES
Cognitive skill instruments,
category classification, 26-28
Educational Testing Service (ETS), 23
Mental Measurement yearbooks, 23
Collaboration,